T0135094

Studies in Computational Intelligence

Volume 528

Series editor

Janusz Kacprzyk, Polish Academy of Sciences, Warsaw, Poland
e-mail: kacprzyk@ibspan.waw.pl

For further volumes:
http://www.springer.com/series/7092

About this Series

The series "Studies in Computational Intelligence" (SCI) publishes new developments and advances in the various areas of computational intelligence—quickly and with a high quality. The intent is to cover the theory, applications, and design methods of computational intelligence, as embedded in the fields of engineering, computer science, physics and life sciences, as well as the methodologies behind them. The series contains monographs, lecture notes and edited volumes in computational intelligence spanning the areas of neural networks, connectionist systems, genetic algorithms, evolutionary computation, artificial intelligence, cellular automata, self-organizing systems, soft computing, fuzzy systems, and hybrid intelligent systems. Of particular value to both the contributors and the readership are the short publication timeframe and the world-wide distribution, which enable both wide and rapid dissemination of research output.

Mirjana Ivanović · Lakhmi C. Jain
Editors

E-Learning Paradigms and Applications

Agent-based Approach

 Springer

Editors
Mirjana Ivanović
Department of Mathematics and Informatics
University of Novi Sad
Novi Sad
Serbia

Lakhmi C. Jain
Faculty of Education, Science, Technology
  and Mathematics
University of Canberra
Canberra
Australia

ISSN 1860-949X   ISSN 1860-9503 (electronic)
ISBN 978-3-662-50946-3   ISBN 978-3-642-41965-2 (eBook)
DOI 10.1007/978-3-642-41965-2
Springer Heidelberg New York Dordrecht London

Printed on acid-free paper

Springer is part of Springer Science+Business Media (www.springer.com)

To my precious Jakov, Tara and Zoran

Mirjana Ivanović

Foreword

Software agents are defined as computer systems situated in an environment and that are able to achieve their objectives by: (i) acting autonomously, i.e., by deciding themselves what to do, and (ii) being sociable, i.e., by interacting with other software agents.

The research and development of software agent technologies has seen a continuous evolution during the last two decades, since they were first proposed as a breakthrough computing paradigm. Especially during the last decade we have witnessed an increased interest in the application of software agent technologies in the context of Internet everywhere and later on Internet on anything endeavors.

Autonomous Agents and Multi-agent Systems represent a well-established research community within the broad area of Computer Science. Research interests in agent systems are spanning various topics including the modeling, design, and development of advanced software systems. Agents have a number of interesting properties including autonomy, reactivity, intentionality, and interactivity that are appealing for a number of contemporary computer applications.

E-learning is broadly understood as the application of digital technologies in education. It represents a collection of e-services that employ digital media and ICT for supporting educational processes. E-learning is a broad term, as well as a highly interdisciplinary area, encompassing many aspects of the educational technologies that cover instruction, training, teaching, learning, pedagogy, communication, and collaboration.

During the last decade, agent technologies were proposed to enhance e-learning systems across at least two dimensions: (i) agents as a modeling and design paradigm for advanced human–computer interaction and (ii) agents for smart functional decomposition of complex systems.

First, agents have been described as entities that exhibit several interesting properties that are very appealing for the modeling and design of advanced user interfaces that are able to adequately support the types of users encountered in e-learning systems: teachers, tutors, and students. The focus of many research works was set on defining various types of adaptation and personalization according to the user type and characteristics.

Second, generic agent types including task agents (i.e., dedicated agents for performing special tasks) and middle-agents (i.e., agents that intermediate between producers or providers of resources and consumers or requesters of resources) were proven to be effective for the appropriate functional decomposition of e-learning systems. Moreover, dynamic and interoperability characteristics of agents are very suitable for supporting maintainability and extensibility of e-learning systems (i.e., agent-based modules are easy to add for extending an e-learning system). For example, specialized task agents can be used for harvesting and composition of e-learning resources according to the teaching, as well as learning needs. Moreover, middle-agents, for example brokers and recommenders, can mediate between the user preferences and resource capabilities, according to different criteria, for example the knowledge level of the student or the educational domain of the course.

Finally, e-learning systems often comprise personal agents that support users during their activities of educational processes. Personal agents can be seen as either suitable for modeling users, their requirements, expertise, and preferences, as well as suitable for implementing advanced modularity concepts that support adaptation and continuous improvement of an e-learning system.

We do believe that e-learning and software agents are orthogonal technologies and their integration is beneficial for designing better e-learning systems. On the other hand, we do expect that this integration will raise new issues and challenges for the future scientific research in this field.

August 2013   Costin Bădică
                                                                        Zoran Budimac

Preface

Over the years, the usage of e-learning and online tutoring systems has proven to bring numerous benefits and advantages over the traditional teaching methods. Recent scientific research has been focused on improving the overall performance of these systems even further. That is, the research focus has been on designing e-learning architectures that exhibit a higher level of adaptability to students personal needs, architectures that offer content personalization, improved motivational aspects, and, in the end, better learning outcomes. During this quest, the increased usage of software agents has shown to be of great importance and challenge. Many concepts of the agent technology, such as intelligence, autonomy, and cooperation, have had a direct positive impact on many of the aforementioned requests imposed on modern e-learning systems.

This book presents the state of the art of e-learning and tutoring systems, and discusses their capabilities and benefits that stem from integrating software agents. We hope that the presented work will be of great use to our colleagues and researchers interested in the e-learning and agent technology.

Synopsis of Book Chapters

The book comprises nine chapters. In Chap. 1, Burkhard, H.-D. and Domańska, M. present their RoboNewbie framework, which simplifies programming of virtual soccer playing robots. The goal of RoboCup soccer leagues is to foster the development of various AI techniques, by providing a formidable challenge in a fun environment. However, the programming of real robots can represent a very complex task. RoboNewbie consists of several easy-to-use APIs that hide these complexities and enable students with no previous knowledge of robotics to quickly start developing their own agents.

Chapter 2 by Kuk, K. et al. takes the similar approach of using games to improve the quality of the learning process. The authors have developed a system of game-based modules for teaching computer science courses. To evaluate the students performance and allow him/her to advance to the next gaming/learning level, their system utilizes a fuzzy logic-based intelligent agent. It has been shown that the developed agent model can be successfully used to assess the students knowledge level, not only in this, but also in various other game-based e-learning scenarios.

Chapter 3 by Gheorgiul, D. et al. presents an e-learning system that aims to preserve cultural identities and heritage of villages in Southeast Europe. As in the previous two chapters, it utilizes the learning-by-playing method. The system combines mobile devices and augmented reality to deliver educational content in the form of virtual reconstructions of environments and objects. Software agents, on the other hand, are employed for the development of narrative e-learning tools, and for the evaluation of acquired knowledge.

In Chap. 4, Tibaut, A. et al. have recognized that, although the virtual learning environments have a significant positive impact on teaching and learning, there is an overall lack of inter-university cooperation. This situation exists mostly due to the lack of interoperability of heterogeneous learning environments. In order to overcome this problem, the authors present a use case of ITC-Euromaster, and propose taxonomy, ontology, and an agent-based software system that enables dynamic interuniversity cooperation.

The next chapter (Chap. 5, by Roy, S. et al.) discusses how agents can be used to manage a grid-based e-learning framework. The proposed system utilizes a number of autonomous, cooperative agents, each with a predefined functionality and responsibility. A concrete implementation of an e-learning architecture as grid services is presented, with the overall conclusion that the usage of agents results in a flexible, convenient, cost-effective, and adaptable framework.

In Chap. 6, Mabanza, N. and de Wett, L. acknowledge that the computer illiteracy represents a major issue in developing countries. Their goal is to evaluate the efficiency of using pedagogical agent in assisting learners to acquire basic computer skills. In the experiment, 103 adult learners undertook a computer literacy training course, with some of them being introduced to pedagogical agent, while others were thought using traditional teaching methods only. The end results have shown that the learners who used pedagogical agents acquired more knowledge and performed better during testing.

The process of finding high-quality learning resources that would satisfy the users needs in a particular context represents a difficult task for any e-learning system. In order help both instructors and learners, Moise, G. et al. (Chap. 7) propose a multi-agent system capable of evaluating and classifying learning material available in open educational repositories. The system relies on the authors socio-constructivist quality model in order to evaluate the quality and relevance of collected material.

In Chap. 8, Gusev, M. presents a novel approach of designing an e-learning system as a cloud service, with an SOA-based architecture. The system operates as an assessment tool, leading the student toward new knowledge by systematically asking questions. To achieve adaptive testing, it employs a number of software agents, each with a different behavior. Evaluation results have shown that the behavior three correct answers in the row results in the best outcome.

In the final Chap. 9, Mušić, D. discusses the importance of emotions, personality, and mood in business-oriented group decision-making processes. The end goal is to develop agents that will be able to successfully assist, or even replace their human users in these domains. The chapter presents research results of implementing emotional agents in e-learning environments. The emotional feature of an agent is constructed using mechanisms and algorithms that simulate experience and patience.

We would like to express our special thanks to professors Zoran Budimac and Costin Bădică for the discussions, constant support and ideas, and their very unique foreword. We also would like to express our thanks to young colleagues Dejan Mitrović and Miloš Savić for technical support.

This book would not have existed without the tremendous contribution by the authors and the reviewers. We remain grateful.

Finally, thanks are also due to Springer-Verlag for their excellent support during the preparation of the manuscript.

Mirjana Ivanović
Lakhmi C. Jain

Contents

**7   MASECO: A Multi-agent System for Evaluation
and Classification of OERs and OCW Based
on Quality Criteria** 185
Gabriela Moise, Monica Vladoiu and Zoran Constantinescu

**8   E-Assessment Systems and Online Learning
with Adaptive Testing** 229
Marjan Gusev and Goce Armenski

Chapter 1
RoboNewbie: A Framework for Experiments with Simulated Humanoid Robots

Monika Domańska and Hans-Dieter Burkhard

Abstract The RoboNewbie project is a basic framework for experimenting with simulated humanoid robots. It serves as an inspiration with introductory experiments for beginners, and it provides room for further challenging experiments. While real robots of this complexity need large efforts for usage and maintaining, the simulated robots run on common computers (e.g. laptops) after short time of installation. They run in the environment of SimSpark RCSS, the official RoboCup 3D simulator for simulated soccer playing robots. The soccer scenario of RoboCup and the related competitions are a well known initiative to foster Artificial Intelligence and Robotics using a popular setting. The simulated robots are models of the humanoid Robot NAO of the French Company Aldebaran. The RoboNewbie framework provides easily understandable interfaces to simulated sensors and effectors of the robot as well as simple control structures. They are illustrated by example agents which can easily be understood and modified by the users. The framework has been successfully used and tested at different courses where the participants needed only few hours to understand the usage of the framework and to develop own agents for different tasks.

Keywords Robotics tutorials · E-learning · Simulated robots · RoboCup

1.1 Introduction

People feel that robots are exciting, others think that they are scaring as in numerous science fiction stories. Are they machines or could they be individuals with own interests and emotions? Will they become competitors of mankind, can

M. Domańska · H.-D. Burkhard (✉)
Institute of Informatics, Humboldt University Berlin, Berlin, Germany
e-mail: nao-team@informatik.hu-berlin.de
URL: http://www.naoteamhumboldt.de

M. Ivanović and L. C. Jain (eds.), *E-Learning Paradigms and Applications*,
Studies in Computational Intelligence 528, DOI: 10.1007/978-3-642-41965-2_1,
© Springer-Verlag Berlin Heidelberg 2014

they make humans to slaves or will they make our life easier? Will they steal our jobs or will they make us free for more human like tasks? Will cold machines substitute human care or will they assist people in managing their daily life? All these questions have to be discussed by politicians, economists, philosophers, moralists etc., but it is difficult to discuss and to solve problems which are superimposed by unscientific impressions.

Thus it is important to teach people more about the techniques of robots and autonomous machines. Such techniques are already integrated in many ways into our daily life, but not as spectacular as robots would be. Of course, mastering these techniques is also an unavoidable prerequisite of economic success today and in the future.

Robotics integrates many disciplines and it is interesting from many points of view. It can provide us with deeper insights not only into (artificial) intelligence, but also into many other fields like physics, biology, medicine, psychology, and philosophy. Robotics is well suited as a comprehensible illustration for many fields in schools. It allows for a broad variety of own activities for building robots and experimenting with them. Related initiatives and competitions are supported by big companies as well as by governmental institutions especially to foster interest in natural and technical science among young people.

It needs not much efforts to build primitive machines which can follow a line, pursue a light, or avoid an obstacle. Children can get first experiences with self-controlled machines. Other tools allow more flexibility and programmed behaviour. More advanced tools are available for university students, a collection of examples is presented in [10]. The RoboCup Initiative [12] uses soccer playing robots of different kinds for development and competitive evaluations of new technologies in Artificial Intelligence and Robotics (cf. Sect. 1.3). The DARPA challenges [4] are attempts to develop new technologies, in previous years for autonomous vehicles, recently for robots acting in a disaster area. Millions of dollars are spent for price money and for governmental support, and only few institutions are able to carry out the necessary efforts.

The main problem that ambitious approaches are confronted with, is the complexity of robots with substantial equipment of sensors and actuators, especially robots with a humanoid shape. Such robots are expensive and need much efforts for their maintenance. In the consequence, most robots used for educational purposes are of simple design. Beyond question, the study of such robots can tell us a lot of the principles of life, and they are able to perform striking tasks. But their intelligence is not sufficient for more complex behavior, e.g. for a robot performing different tasks in a household.

It is the aim of our RoboNewbie project, to provide more ambitious experiments without an explosion in cost and maintenance efforts. Simulated robots in simulated environments are used as an alternative for complex hardware. Our RoboNewbie framework addresses the needs of people interested in Robotics, but it does not assume neither special resources nor specialized education. A simple laptop at home is sufficient such that e-learning will be not restricted by available hardware. Interest in technical problems and only some basic skills in programming (JAVA) are required for the usage, such that even interested young people at secondary schools can use it. It needs only short time for down load and setup, and then the users can immediately start guided by the supporting documents. The framework includes some examples which can be modified and extended such that users can learn by doing. It allows insights to the problems of perception, actuation, and control by related experiments.

We use the existing simulation environment SimSpark. It was developed by the RoboCup Community. It is open source and it is used for the RoboCup robot soccer competitions in the 3D simulation league (cf. Sect. 1.3). It simulates a soccer field with simulated humanoid robots Nao as soccer players. The original robot Nao (Fig. 1.1) is produced by the French company Aldebaran [1], it is also used for educational and scientific purposes. The chapter is continued with some discussions about robots in education. The aims of our RoboNewbie project are explained in more details. Section 1.3 informs about the challenges of the soccer playing robots in RoboCup. Section 1.4 gives a short overview about the SimSpark simulator. The communication between the agents and the simulator is described in Sect. 1.5. Sections 1.6 and 1.7 give an introduction to the RoboNewbie project and its resources for download, while Sect. 1.8 discusses the details of the RoboNewbie framework. Prepared examples and exercises serve as practical introduction to the problems of Robotics and to the usage of RoboNewbie. They are presented in Sect. 1.9. We have evaluated the framework at different courses under different conditions, the results are presented in Sect. 1.10 before we come to our conclusions.

We are thankful to the whole RoboCup community, especially to the developers of SimSpark, to the teams magmaOffenburg and NaoTeam Humboldt, and especially to Yuan Xu. We thank them all for providing the interesting scenario, the resources for simulation, the fruitful discussions and their help. Special thanks goes to the team magmaOffenburg [7]. We have used part of their code for the

Fig. 1.1 Robot NAO from Aldebaran [1]

communication with the server and for the parsing of messages (details are annotated in our code). We also thank the participants of our courses for their engagement and for their evaluations.

1.2 Robots in Education

1.2.1 Experimenting with Hardware

Experiencing with own experiments is an important prerequisite for studies in Robotics and Artificial Intelligence. But experimenting with complex real robots is difficult not only because of expensive hardware. Maintaining the robots and set ups for experiments are very time consuming even for experienced people. Experiments at home as needed for e-learning require a deep technical understanding by the students, i.e. experiences that they are just going to learn. So it is not surprising that simple hardware is still broadly used in Robotics education, hardware which is far behind the recent technical developments, not to talk about complex humanoid robots. The collection of papers in [10] can be understood as an illustration of our statements.

Only few and simple hardware is needed to build robots that can avoid obstacles, follow a line, or drive to a light, respectively. The "Cybernetic Turtle" [21] requires only light-sensitive sensors and separately driven wheels. The sensors and the drives of the left turtle in Fig. 1.2 are connected via cross: If the left sensor gets more light, then the right wheel moves faster and the turtle moves to the left. If the right sensor is closer to the light, the vehicle moves accordingly to the right. Overall, the machine always moves towards the light. If the light moves, the vehicle will pursue it. Obstacle avoiding robots can be build similarly with range sensors on the front instead of light-sensitive sensors. If the light-sensitive sensors and motors are connected left to left and right to right as in the right turtle

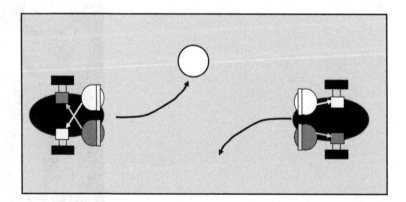

Fig. 1.2 "Cybernetic Turtles": the *left* one pursuits the light, while the *right* one flees from the light

of Fig. 1.2, then the turtle will move away from the lights. Now several turtles of both kinds, some of them with lights on their backs, can be placed together. What happens looks quite complex: Some turtles seem to like each other (those with crossed couplings and lights on the back), others (with parallel connections) seem to be afraid from light carrying turtles and so on. Human observers tend to ascribe even further mental capabilities like tactics and learning, or affections like altruism, brotherhood, revenge. Some primitive intelligence seems to rule the behavior, while only a suitable coupling of sensors and actuators is used. Nature has many examples of organisms that realize very complex behavior in a simple manner. Primitive behavior can be built following related principles [3, 22].

Experiments with such robots can show different aspects of Robotics, among them:

- Basic understanding of sensors, actuators and their coupling,
- Emergence of behavior by embodiment as in Behavioral Robotics,
- Biologically inspired constructions,
- Interaction of primitive individuals as in Swarm Intelligence.

Many available tools and tutorials for basic experiences with robots start with experiments like Cybernetic Turtles and extend it accordingly. The limits of these approaches are reached when more complex situations have to be analyzed and related behaviors are required. Motion in complex terrain and manipulation of objects need more sophisticated actuators (legs, grippers, ...), more sensitive perception (camera with image processing, range finders, monitoring of the internal state: proprioception, ...), and more powerful controls (interpretation of situations, deliberation, monitoring of behavior, ...). Again, there are a lot of robots, hardware packages, sophisticated modules etc. for different purposes available on the market with prices up to several thousands Euro and even more.

From our point of view, humanoid robots are especially appealing for educational purposes. Humanoid robots need to combine complex "low level" functionalities with "high level" deliberation skills. The "low level" functionalities concern the basic bodily skills, which are very difficult to implement in artefacts. Researchers in Artificial Intelligence (AI) had to learn that it is comparably easy to implement journey planners, electronic market places, chess programs, computer games etc. But daily tasks like observing the environment by looking around, fast running, powerful striking, soccer playing, bicycling etc. are much more difficult for machines and offer many still unsolved challenges.

In fact, bodily tasks need much more than forces and muscles, they need complex controls which actually must be trained over long times as it is the case e.g. for dancers, athletes, surgeons, craftspeople. At the same time, the performance of bodily tasks is shaped by the design, by good interactions between the physics of the body, the environment, and the controls, such that optimal behavior can emerge by their interplay. Recent approaches in AI and Robotics claim that human rational behavior is grounded in the interaction with the real world: Intelligence needs a body to grasp the world [11]. For that, experimenting with

Fig. 1.3 Humanoid robots at the RoboCup world championship 2006 in Bremen

humanoid robots leads directly to the challenging problems of understanding (artificial) intelligence and (artificial) life. The interest in robots with human shape is also culturally grounded and there are a lot of (controversial) discussions over centuries. But only recently, the technical developments seems to be mature enough for serious efforts in realization. To discuss about possible consequences, people need to know more about the techniques behind. Such knowledge can be developed by experiments with robots, too. Groups of experienced people can build humanoid robots using customized modules and toolkits (e.g. [14]). About thousand Euros are needed for the hardware to build a robot of e.g. 60 cm height. Figure 1.3 shows humanoid robots of related size at the RoboCup World Championship 2006 in Bremen. The robots are usually equipped with about 20 motor activated joints, a camera and other sensors, and a control board.

Besides such toolkits one can bye well designed complete robots which are available for more than 10,000 Euro like the robot Nao (1) from Aldebaran.

It comes as no surprise, that such sophisticated hardware and the related control software needs some training at the beginning. Moreover, it also needs continuous efforts for maintenance. Together with the reasonable cost price, these efforts often exceed the available resources for educational experiments at schools or at home for e-learning.

1.2.2 Experimenting with Simulated Robots

An alternative is the usage of simulated robots in simulated environments. To be close to real robots, the simulation must model a physical world with rigid bodies, with forces, inertia etc. The robot hardware with the body parts, the sensors and actuators is part of the simulated world, too. Because simulation of real environments is not in the center of robotic experiments, such a simulation should be provided. Then the students can do their experiments by programming robots in this environment similarly to the programming of real robots. For that, the simulated robots must be usable just like robotic hardware in reality, i.e. the students should have access to the sensor data and to the actuator commands similarly to the situation with real robots.

There exist different simulation tools where scenarios and robots can be configured for a physical simulation, usually the systems include already prepared scenarios. Some of them like Gazebo [5] are related to games, others are related to special hardware like NAOsim [2] (for the robot NAO by Aldebaran).

The Robot Operating System (ROS) provides libraries and tools to help software developers create robot applications. It provides hardware abstraction, device drivers, libraries, visualizers, message-passing, package management, and more [15]. Commercially developed simulators (with some free offers for academic purposes) are Webots by Cyberbotics Ltd. [23] and the Microsoft Robotics Developer Studio (MDRS) [9]. They mentioned system contain models of existing hardware and allow the transfer of the code to the real robots. They are in use by many Robotics courses at universities worldwide. The courses using ROS or MDRS assume (or provide) related pre-knowledge for using the systems. The RoboCup community uses several simulators for their competitions (cf. Sect. 1.3 for more details about RoboCup):

- Unified System for Automation and Robot Simulation (USARSim) is a high-fidelity simulation of robots and environments based on the Unreal Tournament game engine. It is intended as a general purpose research tool with applications ranging from human computer interfaces to behavior generation for groups of heterogeneous robots. In addition to research applications, USARSim is the basis for the RoboCup rescue virtual robot competition as well as the IEEE Virtual Manufacturing Automation Competition [20].
- RoboCup 2D Soccer Simulation League is a research and educational tool for multi-agent systems and artificial intelligence. It enables for two teams of 11 simulated autonomous robotic players to play soccer (football) [17]. The robots are modelled by circles in a two dimensional world. There is no physical simulation which makes body control simple, but allows for more dedicated Multi Agent and Machine Learning experiments.
- SimSpark is a generic physical multi agent simulator system for agents in three-dimensional environments. It builds on the flexible Spark application framework. It is used as the official Robocup 3D simulation server [16] for the RoboCup competitions (Fig. 1.4). The robots in the Robocup 3D simulation server are models of the robot Nao by Aldebaran [1].

The RoboCup simulation software is open source which makes it especially attractive for Robotic courses. An excellent example is the course "Autonomous Multiagent Systems" at the University of Texas [19]. For our purposes, the 3D

Fig. 1.4 RoboCup 3D simulation league: finale at RoboCup championship 2012 in Mexico

simulator is especially suited because it is close to real human robot hardware. It is used for educational purposes especially by the universities which have teams participating in the competitions. Several teams have published their source code on the web (cf. e.g. [16], but also the team pages), such that other courses can use it, too. Available tutorial and programming materials are mostly addressing the interests of new team members, and they still need some efforts to become familiar with the server and the soccer agent programs.

1.2.3 Experimenting with RoboNewbie

There exist many materials (robots, simulations, courses) addressing the needs of children on the one side, and the education of undergraduate and graduate students on the other side: The tools for children allow experiments with limited complexity, and the courses at universities need more pre-knowledge and longer engagements. Hence there is some gap in between which concerns interested young people e.g. on secondary schools. We assume that they have some programming skills and technical understanding, but they do not have the knowledge provided in computer science curricula.

The charm of products like Lego Mindstorms [6] is the fact, that people can immediately start without much preparation. They can learn by doing and incrementally step forwards to more complicated tasks. It the aim of our RoboNewbie project to provide a fast and easy entrance to the world of humanoid robots. Simple experiments lead to a basic understanding of the problems in perception, actuation, and control, respectively. There are no prior requirements besides some programming skills and interest in technical problems. At the same time, there are no limits for step wise extended tasks which require more knowledge as it can be provided by courses in Robotics. This makes RoboNewbie also a good candidate for practising at higher level education (cf. Sect. 1.10).

We have chosen the SimSpark simulator of the RoboCup community as the environment of a our simulated robots for several reasons: It is well established and used by a broad community, it is open source and can be modified without problems. It has minimal requirements for resources and can run on simple hardware (e.g. on laptops). It was developed and is still maintained by volunteers of the RoboCup community and hence improved and tested over years. It opens close connections to the world wide RoboCup community with the option to participate later at the RoboCup competitions.

The minimal assumptions for the usage of RoboNewbie can be provided even by a technically oriented education at secondary schools. Hence the framework could be used already on this level:

- Some basic programming skills for JAVA are required such that users can write their own programs. All robot related processing in RoboNewbie is transparent to the users in detail, we did not want to hide details e.g. by simply clicking graphical icons, or vice versa by complicated program structures or unexplained algorithms. The documentation in the provided code explains the methods and the conditions of their usage.
- Some mathematical and technical understanding is assumed to understand the methods involved. The examples for beginners do not require deep understanding of Robotics. Instead, they lead to a first understanding of the problems to be solved. Education in Robotics can use the framework from the very beginning and later come to more complicated tasks.

The RoboNewbie framework provides the necessary program structures which simulate the middle ware of a robot (access to sensors and actuators, agent structure, connection to the server, etc.) and the parameters of the simulated world (size of objects, identifiers, etc.). Several examples give introductions to the usage of the framework and to the problems of Robotics. First exercises require only modifications of the examples, such that successful own work can be experienced after short time. Besides the programs, some documents explain the usage of the framework and guide the experiments and exercises (cf. Sect. 1.10).

At the recent stage of development, the materials presented at the RoboNewbie webpage [13] are ready for usage as e-learning material associated to introductory courses on Robotics. We have successfully tested the usage at several short courses at different universities (cf. Sect. 1.10). Based on these experiences, the lecture materials for providing theory and experiments in a unified style for e-learning are in preparation.

The usage of a soccer simulation in the RoboNewbie project has several advantages: The robots act in a dynamically changing environment with given structures and rules, which are well known to the public. To play soccer the robots need appropriate skills for motions, they need perception to observe and interpret the situation, and they need controls to react in real time. They have to coordinate with their team members, and they have to be aware of their opponents. The scenario allows for competitions between the users which serves for motivation. Experiments can range from the development of simple motion skills up to challenging tasks in Machine Learning.

1.3 Soccer Playing Robots: RoboCup

The RoboCup community has 15 years of experience with real and simulated robots in the field of soccer playing robots [12]. Acting autonomously in the real world is still difficult for machines. Therefore, the development of soccer playing robots has become a challenging research and test field. The competitions in RoboCup are used to evaluate scientific and technological progress, similarly to the role of chess as a test field for Artificial Intelligence in the past. Soccer playing robots have to be able to control their bodies and their motions according to soccer play, they must perceive a dynamically changing environment and they have to choose successful actions out of many options in real time. They have to cooperate with team mates and to pay attention to opponents. Several thousand scientists and students are participating in the annual RoboCup competitions in different leagues with different types of real and simulated robots. The different leagues were introduced to tackle different problems based on different hard- and software. Some of them can drive on wheels, while others have humanoid shapes with two legs.

The humanoid robot Nao is used in the Standard Platform League (SPL), where all teams use the same standardized hardware. Its simulated version is used in the 3D-simulation league. The official SimSpark RoboCup 3D Soccer Simulation (SimSpark RCSS) [16] provides an excellent environment for experiments with simulated complex robots: It provides a physical simulation using ODE [18] for the body dynamics of the robot Nao and the soccer environment. Users can program their own robot controls as "agents" which communicate by messages with the simulation server of SimSpark RCSS. The agents could be considered as the "brains" of the robots. They perceive sensory information from the server and send action commands back to control the motors of the robot. The technical details of the simulation server are designed for the more experienced users participating in RoboCup. They can still be a barrier for inexperienced users and some efforts are necessary to become familiar with them. Therefore, our RoboNewbie project addresses especially the needs of such inexperienced users. A more detailed introduction to SimSpark RCSS will be given in Sect. 1.4. The robots in RoboCup have to act autonomously, no human interaction is allowed.

They have to recognize the environment, to decide about their next goals and related actions, and to perform the actions using their skills. The robots have to gather all needed information using their sensors (Fig. 1.5). They have to process the sensory input to obtain a picture about the situation, the localization of the robot itself, of the other robots, and of the ball. Today, visual sensors are widely used to perceive the environment. Sophisticated algorithms for picture processing and scene interpretation are needed. Statistical methods like Kalman filters or particle filters are used for localization tasks. Not only the place but also the the direction and the speed of the ball are very important. Latency modeling and prediction methods are important as well.

Especially humanoid robots need various proprioceptive sensors for observing and controlling their movements. Sensors for joint angles, forces, and torques measure the positions, directions and movements of different parts of the body.

Having a belief (not necessarily a true knowledge) about the environment, the robot has to decide for its next goals and actions. This means to check and to evaluate the own chances in comparison with the opportunities of other robots (team mates and opponents) on the playground. Therefore, the robot needs knowledge about his own skills and about the results it can hopefully achieve.

There are different levels of control. On the lowest level, the robot has to control its body movements. In the case of humanoid robots it has to keep balance while walking or kicking. This needs a continuous interaction between sensor inputs and appropriate actions at the related joints. The compensation of an unexpected force by an adjustment of the heap is an example. It is still an open problem in the worldwide research on humanoid robots how this can be achieved best: how to couple sensors and actors, which sensors to use, how to program the control etc. Recent efforts try to implement some kind of a spinal cord inspired by solutions from nature. Because of the lack of complete models, methods from Machine Learning are tested for the development of efficient (distributed) sensor-actor loops.

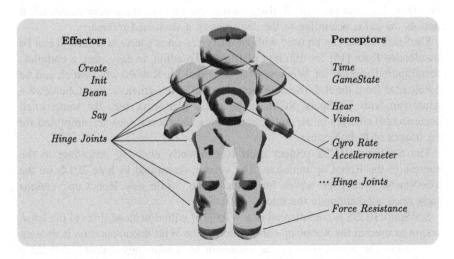

Fig. 1.5 The effectors and perceptors of the Nao in the simulator SimSpark RCSS

Having such basic skills like dribbling, intercepting or kicking the ball, the next level of control concerns the choice of an appropriate skill for a given task. While the skill is performed, the robot has continuously to check the performance of the skill, e.g. maintaining control over the ball while dribbling. Again, a close inter-action is necessary between sensors, control, and actors.

On the highest level(s), tactical and strategic decisions can take place. Actually, related reasoning procedures are especially studied in the simulation leagues (preferably in the 2D simulation). Psychologically inspired attitudes like belief, goals, plans, utilities etc. are used.

1.4 SimSpark RoboCup 3D Soccer Simulation

The SimSpark RoboCup 3D Soccer Simulation (SimSpark RCSS) is developed and used by the RoboCup community in the 3D Simulation League. It simulates the soccer games in a three-dimensional world, while the games of the 2D Sim-ulation League are restricted to only two dimensions without complex physical features. Hence, the 2D Simulation is especially used for multi-agent modelling, while the 3-Simulation League models soccer play regarding the physical prop-erties of the real world.

SimSpark is a generic physical multi agent simulator system for agents in 3D environments. It uses the Open Dynamics Engine (ODE [18]) for detecting col-lisions and for simulating rigid body dynamics. ODE allows accurate simulation of the physical properties of objects such as velocity, inertia and friction.

The Simulator SimSpark RCSS consists of the server for simulation and the monitor for visualization and interaction, together with some configuration files. It models a soccer field with the player bodies (adapted from the robot hardware of Nao) (Fig. 1.4) and the ball. It also controls the rules of the soccer game, i.e. it controls the game according to the decisions of a simulated referee.

SimSpark RCSS can be used without charge as open source software. It can be downloaded from [16] for different platforms. To admit an easy start, a complete preconfigured version for Windows 7 is provided for RoboNewbie which can be downloaded from the RoboNewbie web page [13]. Nevertheless, the RoboNewbie agents run with SimSpark RCSS under other platforms, too. By some small (documented) changes in the configuration files, the soccer rules are simplified for first usages of RoboNewbie.

The SimSpark RCSS project itself is constantly evolving according to the progress in the RoboCup initiative. The version (compiled in June 2012) on the RoboNewbie web page serves for stable usage, while new RoboCup versions might need adaptations in the future.

SimSpark RCSS is documented in a Wiki [16] with download links to the latest version as used in the RoboCup competitions. The Wiki documentation is thought to represent the actual state of the simulator by continuous updates. But since different developers are volunteering in parallel on different tasks in the project, the structure of the Wiki is not always optimal, and occasionally some outdated information is still present. Moreover, the Wiki is directed to experienced users which makes it sometimes difficult to understand for novices.

To provide an easy access, the down loads of the RoboNewbie Project contain an introduction to SimSpark RCSS which refers to the provided version (as described above). It gives the user an overview about

- Simulation using SimSpark RCSS: The SoccerServer and the Monitor,
- The Nao-Model used by SimSpark RCSS,
- Communication between agents and SimSpark RCSS (with explanations of the message formats),
- Synchronization between SimSpark RCSS and the agents,
- Monitor and user interface,
- Running a game.

Actually, our description of SimSpark RCSS provides also some "background" information which is not needed for beginners, e.g. details about the message formats. Since RoboNewbie permits a direct access to the items of messages like sensor values and motor commands, the syntax of messages must not be known by users. Nevertheless, we have included the information for deeper understanding of RoboNewbie in case of interest.

1.5 Communication Between Agents and SimSpark RCSS

SimSpark RCSS implements the soccer environment including the bodies of the Nao robots. It models all physical interactions between players, ball and environment. The agents implement the control of the players. The interface between the physical environment and the control of real robots is constituted by sensors and actuators: Robots perceive the world by sensory data (e.g. by vision, accelerometer, force sensors etc.), and influence the world by their actuators (motors, voice etc.). In simulation, the sensory data are calculated by the simulator according to the situation in the simulated world (e.g. observable objects) and sent via message exchange to the agent. Then, like a real robot, the agent can update its belief about the situation and decide for actions it wants to perform. A real robot would then activate its actuators (e.g. motors at the joints) to perform the intended actions. The agent communicates with SimSpark RCSS by messages which transmit the sensory data and the motor commands, respectively. Both are synchronized by a communication cycle of 20 ms (Fig. 1.6).

The message transfer with SimSpark RCSS is optimized for minimizing the server load: All sensory data are packed in one server message to be sent at the beginning of a communication cycle. Vice versa, the agent can send all action commands by a single agent message before the end of a cycle. The message formats follow a special syntactic scheme based on symbolic expressions (S-expressions). As a consequence of collecting data into one message, the processing of the data in an agent needs more efforts than in a real robot. It is an advantage of the RoboNewbie agent that this preparation is hidden from the user: The agent provides special getter- and setter-methods which allow the access to the sensor (perceptor) data in a similar way as in a real robot.

The interaction between the server and the agent works as follows (cf. Fig. 1.7):

1. At the beginning of a cycle at a time t, the server sends the individual server messages with sensations to the agents.
2. During this cycle, the agents can decide for new actions depending on their beliefs about the situation.
3. Before the end of this cycle, the agents should send their agent messages to the server for desired actions.
4. The server collects the agents messages and calculates the resulting new situation (poses and locations of the players, ball movement etc.) according to the laws of physics and the rules of the game. This is done during the following cycle at time $t + 1$. (Note that the server message sent at the beginning of this cycle regards the situation calculated in the previous cycle at time t).
5. At the beginning of the subsequent cycle, at time $t + 2$, the sensor data in the server message is based on the effects of the actions at time $t + 1$ which were chosen by the agent according the information from time t.

The delay between observation, reaction, effects and recognition of effects corresponds the situation in the real world. Students learn to regard these effects when programming with RoboNewbie. Another special feature of SimSpark RCSS is the use of so-called perceptors instead of sensors. The perceptor data can be regarded as already pre-processed sensor data. For example, the image data from the camera are not presented by a pixel matrix. Instead, the vision perceptor sends a collection of observable objects with egocentric coordinates relatively to the camera of the observing agent. In a similar way, action commands of the agent are encoded as so-called effector values and sent to the server which translates them to the intended actuator control commands. The calculation of perceptor values and the interpretation of effector values are part of the simulator, too (see Fig. 1.6). On the agent side, a server message has to be parsed for the contained perceptor values, and the action commands have to be collected to the agent message. Both

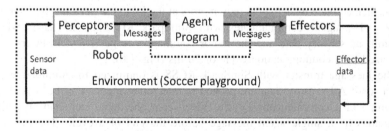

Fig. 1.6 Simulation Scheme: All parts in between the *dotted lines* belong to the simulation of the physical world. They are simulated by the SimSpark RCSS

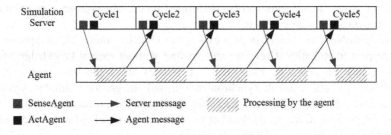

Fig. 1.7 The synchronisation between the server and an agent

constitute a significant burden for a beginner while it provides only few insights into robotics. But as already mentioned, the RoboNewbie users need not to care about that, because the needed processing is done by the framework (cf. Sect. 1.8.1).

The simulated robot has Hinge Joint Effectors for each of the 22 hinge joints and a Say Perceptor (as of a loudspeaker with limited capacity). Besides them, the initial connection with SimSpark RCSS is installed by special effector messages, too. They define the team identity (team name, player number) and the initial position.

The following perceptors are available in SimSpark RCSS (for details see the Wiki or our SimSpark description):

- Vision Perceptor (camera in the center of the head),
- Hinge Joint Perceptors at each of the 22 hinge joints (cf. Fig. 1.11),
- Accelerometer in the centre of the torso,
- GyroRate Perceptor in the centre of the torso,
- Force Resistance Perceptor at each foot,
- Hear Perceptor (directed microphone with limited capacity),
- Game State Perceptor (reports the actual game state of the soccer match).

1.6 The RoboNewbie Project

The RoboNewbie Project is a basic framework based on JAVA for the development of simulated humanoid robots. It provides easy understandable interfaces to simulated sensors and effectors of the robot as well as a simple control structure. It serves as an inspiration for beginners and behinds that it provides room for many challenging experiments. It runs in the environment of the SimSpark RCSS, thus it can but need not be used for soccer playing robots. Users can develop their own motions, e.g. for dancing, gymnastics or kicking a ball. They can get insights into the complex phenomena of coordinated limb control, of kinematics and sensor-actor control. They can experiment with problems of perception, action planning, and coordination with other robots. The framework can also be used for Machine

Learning, where many runs can be performed to train behaviours—much more than ever possible with real robots.

The RoboNewbie Project implements some kind of "minimalistic approach" with respect to Robotics. Users are able to start without special knowledge about robots. They can learn by their own experiences about the basic concepts of perception, motion, control, synchronization, and integration. Simple exercises lead to more insights. The first exercises are based on prepared examples and described by tutorial materials. Further exercises may be solved in parallel while presenting more theory in a Robotics course.

All related program code in RoboNewbie is understandable from simple principles without further knowledge. That concerns the structure of the code as well as the underlying computational methods. As soon as users learn more in Robotics, they will be able to extend the programs accordingly, e.g. concerning complex motions or world modelling.

At the same time, such a "minimalistic approach" is not needed for the non-robotics aspects of RoboNewbie, e.g. the communication with SimSpark RCSS and the special preparation of messages (cf. Sect. 1.5). RoboNewbie performs all this tasks in the background. The users need not to deal with these details. Instead, RoboNewbie provides comfortable access methods. Users need not to be aware that they are dealing with simulated robots instead of real ones.

1.7 The Resources of the RoboNewbie Project

The main goal of the RoboNewbie Project is to provide an uncomplicated starting point to the programming of complex robots with minimal requirements and pre-knowledge. The users are only supposed to have some programming background (Java) and some technical/mathematical understanding. More knowledge about robotics can be provided in parallel to the exercises with RoboNewbie, e.g. as a course (as we already did) or by further elaborated e-learning materials.

The users of the RoboNewbie project can find all materials on the web page [13] of Berlin United/Nao-Team Humboldt. Besides links to RoboCup, Nao (Aldebaran) and the SimSpark-Wiki, it contains resources for download:

- Description of installation and first steps by the documents *Installation* and *HowToStart*.
- Sources of the RoboNewbie Agent programmed in JAVA 7 and prepared for usage under Netbeans.
- *Quick Start Tutorial*: Introduction to the features and the usage of the agent.
- *Motion Editor* for the design of Keyframe Motions (needs JAVA 3D to be installed) with an introduction.
- SimSpark RoboCup 3D Soccer Simulation (SimSpark RCSS) for Windows with an introduction to the usage of SimSpark RCSS for RoboNewbie.

1.8 RoboNewbie Framework

The RoboNewbie framework offers a comfortable interface for agents interacting with SimSpark RCSS. It includes sample agents which illustrate basic concepts and methods of Robotics (cf. Sect. 1.9). Users can start exercises with these agents and learn how to use RoboNewbie and what the programming of robots is like. They can make their own experiences with different topics and algorithm by modifications and extensions. The source code of the framework and the agents is open source.

It is a main goal of the project, to provide easily understandable concepts, methods and programs, which need no special education or training to start with.

There are no complicated structures, and all code is documented in detail. As a consequence, some more demanding concepts were replaced by simpler approaches (e.g. keyframe motions instead of inverse kinematics, approximated coordinates of observed objects etc.). Nevertheless, the clear structure of the project supports extensions for more challenging solutions if wanted.

1.8.1 Low Level Interface Functionalities

The framework includes interface functionalities on two levels. The lower one corresponds to the hardware-near functionalities of robots, while the higher one is concerned with more abstract control functionalities, cf. Fig. 1.8. For the

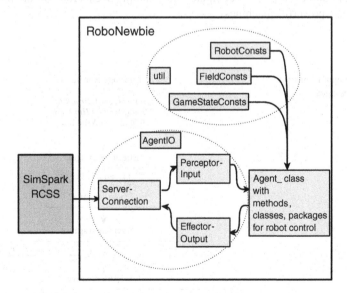

Fig. 1.8 Data flow for RoboNewbie as interface to the simulator

communication with the server, parts of the code of the RoboCup team magma-Offenburg [8] was used as documented in our source files.

The hardware-near layer encapsulates the network protocol for interaction with SimSpark RCSS and it allows access to the simulated hardware entities corresponding to sensors and motors. The access is implemented by getter functions for perceptor values of different perceptors which can be used similar to sensor signal queries of real robots. Related setter functions for effector values can be used for control of actuators. Especially the low level interface functionalities in SimSpark RCSS are a hurdle for beginners, and they need substantial work even for experienced users. They concern tasks like network connection, synchronisation with the server, parsing of nested server messages, syntactical analysis of S-expressions, synthesis of agent messages with a lot of technical non-robotics details. The users of RoboNewbie need not to care about that, the framework offers ergonomic methods for the interaction with the simulated environment in an easily understandable way similar to the methods used by the operating systems of real robots. Users can learn to use these methods after a short training time (cf. the evaluation in Sect. 1.10).

The processing of effector and perceptor messages is illustrated by Figs. 1.9 and 1.10, respectively. The synchronization protocol was already described in Sect. 1.5. The user needs not to care about the communication details, except the delays by the protocol and the duration of the cycles given by 20 ms. It is necessary to fetch a server message at each cycle and to send the agent message before the end of the cycle. The related control structures are already implemented in the examples and explained by the *Quick Start Tutorial*. If the calculations during one cycle do not exceed the cycle time, there will be no problem. The needed time depends of course on the used computer, the example agents run without problems even on less powerful machines.

Fig. 1.9 Preparation of effector messages

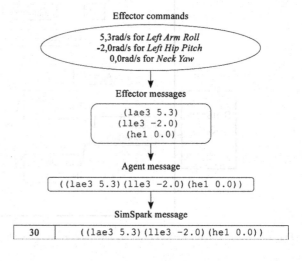

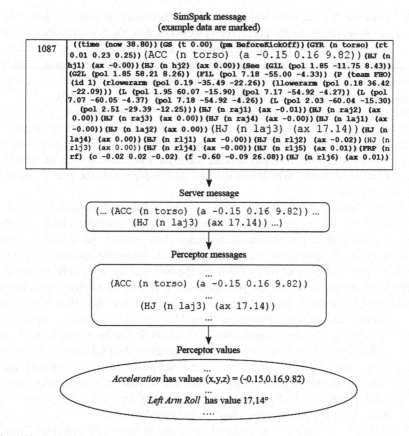

Fig. 1.10 Analysis of perceptor messages

The first example `Agent_BasicStructure` (cf. Sect. 1.9.1) lets the user start with an agent which already implements all low level communication. The agent simply rises an arm by setting related effector values. The user can experiment with other values and other effectors just to understand the basic structures.

1.8.2 Perception

The available perceptors were already listed in Sect. 1.5. All perceptor values can be queried by related getter methods using the perceptor names instead of the acronyms of the server messages. This allows a comfortable access to the perceptor data which corresponds to the access of sensor values by a related operating system of a real robot.

RoboNewbie has already implemented the necessary conversion from the nested server messages to the perceptor values. For that, the S-expressions of the server message are parsed for the constituents of a tree like structure (again, thanks to usage of the code of the team magmaOffenburg [8] as documented in our code). According to the analyzed acronyms in the expressions of the tree, the corresponding perceptor values are filled in by RoboNewbie (cf. Fig. 1.10).

The programs `Agent_TestPerceptorInput` and `Agent_TestLo-calFieldView` (cf. Sect. 1.9) illustrate the usage of the related getter methods and the perceptor values. As an exercise, the user can implement an agent, which lifts the robots arm, when it senses another robot and moves the arm down, when it does not sense any robot. Which arm is lifted should depend on the side where the other robot is seen.

Special efforts are needed for the vision perceptor. It provides coordinates of all objects in the vision range of the camera of the robot. SimSpark RCCS in its common version does not communicate image data. Instead, the communicated information can be understood as the result of basic image interpretation, it contains coordinates of the goal posts, the lines, the ball, and the body parts of robots.

The coordinates of the vision perceptor are given by egocentric coordinates relatively to the camera in the centre of the head. Hence they depend on the position of the head. Further calculations would be necessary to get the coordinates of objects relatively to other coordinates, e.g. relatively the body direction of the robot ("robot coordinates") or to global coordinates of the playground. Accurate calculations would need the inspection of the cinematic chain. The necessary data are available by the hinge joint perceptors and the inertial sensors (accelerometer and gyrorate perceptor). Further calculations especially for self localization would be necessary for the transformation into global coordinates.

RoboNewbie does not provide related programs following the intended "minimalistic" approach, because they would not be understandable by beginners without pre-knowledge about Robotics and Linear Algebra. Instead, the implementation of related methods can serve as exercises during courses in Robotics.

As a simple substitute, we have decided to provide only approximations for the conversion from camera coordinates to robot coordinates. The approximation regards simply the offsets by the turning angles in the program LocalFieldView. It is documented in the sources and easily to understand. It is correct only if the head is turned horizontally, but not for a tilted head. Users can make experiments according to the accuracy and draw their own conclusions on cinematic relations. Some problems due to the approximation are discussed in Sect. 1.10.2.

Visual information is provided by SimSpark RCSS only at each third cycle, and the robot would have to act blindly in between when there are no vision data available. Because of that, the perceived vision information should be stored for the following cycles. Moreover, the vision perceptor is limited by the camera view range of 120° horizontally and vertically. Hence, the robot has to move its head to observe more objects in the world. Again it is useful to store objects seen before in other directions. In general, such updating and memorizing of past and recent observations is maintained as belief of the robot in a so called world model.

Updates may regard corrections according to robot motion, guesses for movements of invisible objects and integration of information communicated by other robots.

Again, a fully elaborated world model is far behind the scope of beginners. Hence, RoboNewbie provides a very simple version, where just the observed objects are stored in a simple form. The coordinates of those objects are referenced with respect to "robots coordinates", where we use the coordinate system of the camera when facing forwards (when neck pitch and neck yaw angles are zero). The correction is done by simple approximate calculations as mentioned above. Other movements of the robot like turning or walking are not regarded. Time stamps indicate the last time of observing an object.

1.8.3 Motions

All intentional motions are performed by controlling the hinge joints by sending effector values (speed of motors) to SimSpark RCSS. Then the physical simulation engine calculates the effects of the commands regarding physical laws and updates the simulated world accordingly. The simulated Nao has 22 actuated hinge joints (cf. Fig. 1.11) which can be controlled by motor commands every 20 ms. It results in 1,100 commands per second.

Simple motions like turning the head or rising the arms can be easily programmed by the users. The motions can be controlled using the feedback of hinge joint perceptors. i.e. by sensor-actor coupling, where the delay of observing an action has to be regarded as described in Sect. 1.5. There is much room for own experiments of users.

More complicated motions like walking need coordinated movements of different joints, users may learn about these problems after some trials. We have decided to provide keyframe motions in RoboNewbie because they are easily to understand and to design. The interpolation mechanism for keyframe motions in RoboNewbie realizes a linear interpolation—users may implement other interpolation methods like splines if they want. Keyframes are stored as text files which can be edited by any text processing system. There with, users could even design and change motions while using the programs as a blackbox.

RoboNewbie comes with a set of very simple predefined keyframe motions for walking, turning, stand up and others. The simplicity is intentional: The examples illustrate only the principles, while the users are encouraged for improvements. They can change these motions by changing the related text files in the keyframe directory. They can also define new motions and integrate them into the framework. Details are explained in the *Quick Start Tutorial* and the documentations of the related programs.

According to simplicity, there are no concepts implemented for interruption of motions: Each motion is performed completely until its end, and there are no cyclic motions, e.g. for walking. Instead, continuous walking is performed by subsequent calls of a two-step-walk.

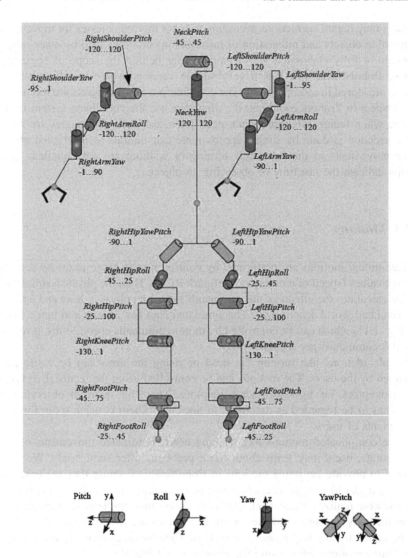

Fig. 1.11 The hinge joints of the Nao in the simulator SimSpark RCSS. Positive turn direction is from x- to y-axis

The design of keyframe motions is supported by the *Motion Editor* which was developed by the NaoTeam Humboldt. It can be downloaded from the Robo-Newbie web page as well. It shows the postures of the robot for selected keyframes. Then the keyframe can be edited in two ways. In the graphical representation the posture can be kneaded into the desired posture with the mouse. Alternatively, each joint angle can be set to specified values which are immediately presented by the graphics. Transitions between keyframes can be defined with specific transition times as explained in the provided documents.

The program `Agent_KeyframeDeveloper` helps in designing keyframes. A robot performs the motion of the actually edited keyframe file. After each change, the new motion is performed immediately. If the robot falls down during such a motion, it stands up by itself. Another helpful program can be used to mirror keyframes from one side to the other.

The example `Agent_SimpleWalkToBall` illustrates the motion concepts (cf. Sect. 1.9). As an exercise, the users can change the program for obstacle avoidance (walk around the ball without touching it). To realize it, they can use existing keyframe motions for walk, stop and turn. Additionally, the agent must be able to recognize the ball and to decide for the appropriate motions. Another exercise is the design of a new motion for kicking the ball. Users can furthermore do their own experiments e.g. with dancing robots.

In general, keyframe motions are useful for designing special motions like standing up, but they are not so well suited e.g. for walking. Walking is still a challenging problem in Robotics. The users of RoboNewbie will get some understanding about these challenges. The framework is also well suited as a basis for other implementations and for Machine Learning by more educated users.

1.8.4 Control Cycle and Decision Making

The basic control cycle follows the classical deliberation approach, often denoted as the "sense–think–act–cycle", or by related similar names. This corresponds closely to the cycle given by SimSpark RCSS: At first, sensations are provided to the agent, then the agent decides for appropriate plans and then it sends the related action commands back to the server.

Critical remarks may come from the community of Embodied Robotics/AI, e.g. concerning the centralistic and symbolic computations in the classical approach. To realize the promising concepts of Embodied Robotics/AI one needs to put more emphasis on local sensor actor coupling, distributed control, embodiment, situatedness, emergent behaviour etc. The real robot Nao as well as its simulated counterpart with their central control (i.e. our agent) are not primarily designed for such purposes. It is possible to design sensor actor couplings and other behavioural concepts in the RoboNewbie framework, too. One might even split the agent into different "parallel" acting parts (implemented e.g. by threads) to simulate distributed controls, but some synchronization is unavoidable by the server cycles of SimSpark RCSS.

At the same time, thinking in terms of the "sense–think–act–cycle" is quite natural for beginners because it reflects some causal dependencies. It provides an intuitive and easily maintainable structure in the design of robots. Therefore, the control cycle in RoboNewbie adopts the related terms for structuring the run-methods of the agents by cyclic calls of methods sense, think and act. The think-method is sometimes omitted for simpler ("reactive") agents.

The sense method is responsible for receiving and processing the perceptor data by the already implemented RoboNewbie methods. The act methods calls the transfer of the agent message with the effector commands. What is left is the further analysis of the perceptor data (e.g. a more elaborated world model) and the decision for plans and actions to be performed now and possibly in the future. By the given structure of RoboNewbie, all this can be included in the think method. The think method can of course be split into more dedicated deliberation methods which may be organized hierarchically if needed. Again, all this is left to exercises during related courses. RoboNewbie provides just a simple example for illustration, the program `Agent_SimpleSoccer` (cf. Sect. 1.9).

The `Agent_SimpleSoccer` is able to perform a very simple soccer play: As long as it is behind the ball and sees the opponent goal, it walks forward while pushing the ball with its feet. If the condition is not fulfilled, it turns around until it sees the ball, walks to the ball, turns around the ball until it sees the opponent's goal, and then it starts walking towards the goal again. The decisions are made by a simple decision tree whenever the previous motion is completed (note that motions can not be interrupted as described above).

The play of `Agent_SimpleSoccer` can be improved in many ways. This is just what we want: The users can collect many ideas for improvements. Improvements may concern better usage of perception (e.g. by a ball model guiding the search), improved motions (like faster walk), new motions (like kick or dribble), better control (like path planning). It is also possible to have more players on the soccer field such that players can cooperate (e.g. by positioning and passing). There with the RoboNewbie agents can be extended for competitions in a course.

1.8.5 Logger

Runtime debugging of programs may be difficult because it affects synchronization with the server: While an agent stops at some break point, the server runs on and the agent program will be not synchronized anymore. Debug messages printed on System.out may need too much time such that the agent cannot respond in time and becomes desynchronized again. It is possible to use the so-called sync mode which lets SimSpark RCSS wait until all agents have sent their messages [16]. Alternatively, all debug messages can be collected by the program `Logger` of RoboNewbie. After the agent has finished, the collected messages are printed out. The usage is shown by the programs `Agent_TestPerceptorInput` and `Agent_TestLocalFieldView` (cf. Sect. 1.9). Both programs provide also examples for the usage of the access methods for perceptors and effectors.

1.9 Exercises

RoboNewbie provides prepared examples for different aspects of Robotics. They contain sample code to be used, modified and extended as exercises. They serve as examples for own implementations. Step by step, the examples help the users to learn more about problems and methods in Robotics and to become familiar with the usage of RoboNewbie. All code is easily understandable, and the detailed documentation explains the methods behind and how they are used.

The examples were already mentioned in the preceding section. All example agents have the same architecture. Identical methods for initialization establish the communication with the server and define team membership and initial pose. The run-methods implement the sense-think-act-cycle. Dedicated methods (or classes for more complex cases) implement sensing, thinking, and acting, respectively.

The *Quick Start Tutorial* provides necessary hints and the links to the examples. It is already a step towards the e-learning material. It will be extended and integrated into a complete online course. In our practical evaluations (see Sect. 1.10.2), the experiments and exercises of the *Quick Start Tutorial* were connected to the lectures.

1.9.1 Hello World: Structure of Agents, Simple Actions

The program Agent_BasicStructure is a first simple example demonstrating the architecture of the RoboNewbie agents (cf. Fig. 1.12). It shows basic concepts of the RoboNewbie framework and gives examples for interacting with the simulation server and using the classes EffectorOutput and PerceptorInput (cf. Fig. 1.8). The usage of the framework is explained by help of this examples and users can try out modifications as exercise 1:

- Choose another initial position for the robot,
- Change the effector commands,
- Try out other velocities,
- Use other robot joints.

The users can understand the usage of motor commands (effectors) under control by joint perceptors values (i.e. sensor-actor coupling). They become able to implement other simple examples, e.g. performing knee bends or dancing. They will understand, that implementation of walking needs more efforts.

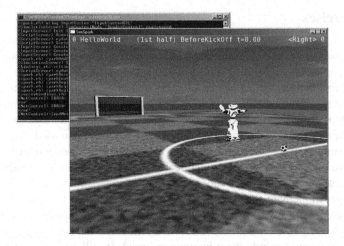

Fig. 1.12 The Hallo World Example

1.9.2 Examples for Perception

The programs Agent_TestPerceptorInput and Agent_TestLocal-
FieldView illustrate the usage of perceptors. Examples for the usage of perceptors and the perceived information at several cycles are collected by the logger
and printed as output. The code provides examples how to get this information
(learning by examples). Agent_TestLocalFieldView helps to understand
the collection and processing of data by looking around (cf. Sect. 1.8.2).

The agent is the base for exercise 2. The task is the implementation of an agent
which signals the direction of an other moving agent. It has to lift an arm when it
senses another robot and moves the arm down, when it does not sense any robot.
Which is lifted depends on the direction: If the other robot is on the left (right)
side, the left (right) arm should be lifted. The example agent Agent_Simple-
Soccer can be used as a moving target.

1.9.3 Examples for Motion

The RoboNewbie project provides already some simple keyframe motions. The
example Agent_SimpleWalkToBall illustrates their usage. It is the base for
exercise 3 where perception and motion have to be combined to avoid some
obstacle. The robot has to start facing the ball. It should walk forward, not hitting
the ball while staying close to the direct way. For that it must move sidewards
when it comes close to the ball, walk past the ball and then turn back to the original
direction. The evading is composed from related keyframe motions.

Exercise 4 concerns the design of keyframe motions and their integration into the framework. The task is the design of a motion for kicking the ball. It can be developed using the *Motion Editor*. The program `Agent_KeyframeDevel-oper` helps for the design: It can run continuously and it performs the new keyframe motion immediately after it was saved as a text file. If the robot falls down, the program performs a stand up motion such that it is ready for a new trial.

The new motion must be integrated into the framework according to the instructions given by the code documentation. Finally, an agent performing the developed kick has to demonstrate the successful implementation.

1.9.4 Examples for Control

The program `Agent_SimpleSoccer` implements a simple soccer player as described in Sect. 1.8.4. It uses all features of RoboNewbie under the control by a simple decision tree. The decisions depend on the situation that is perceived by the perceptors. Other examples of control structures can be found in the examples mentioned before.

The `Agent_SimpleSoccer` is extremely simple: It has no kicking skill, its perception is very raw, and its decisions are not very accurate for the actual situation. It mostly needs about 10 min to push the ball into the opponent goal. Its simplicity comes again by intention: The users of RoboNewbie have many chances to improve the behavior. They can implement a kick motion (exercise 4), and improve the given motions for walk and turn. They can improve the perception such that robot does not loose so much time for orientation, e.g. by a better world model. The control can be improved for shorter reactions and better adaptation to the situation. Improved control parameters are one possibility for that. Users are allowed to use more that one robot, such that they can implement coordination and communication.

We have used this exercise for final competitions with the participants of our introductory courses (cf. Sect. 1.10). It appeared to be a good source for motivation.

While dealing with the exercises, students experience basic principles of Robotics and become able to follow theoretical lectures with better insights and motivation. Further examples and exercises illustrating more advanced methods can be developed according to the needs of related courses.

1.10 Evaluation by Courses

We have tested the RoboNewbie framework under different conditions and with different users. The results are regarded for the further development of the framework and the e-learning project on Robotics. We have used the framework for several courses on Robotics Fig. 1.13 and Fig. 1.14:

Fig. 1.13 Participants of the DAAD course in Ohrid, August 2012

1. Short Seminar on Robotics, University of Novi Sad, Serbia, June, 5th and 7th, 2012, 4 h, 12 participants. This Seminar served for a first test of the concepts.
2. DAAD school on Robotics and Mathematics Ohrid, Macedonia, August, 12–18, 2012, 30 h, 23 participants.
3. Course "Cognitive Robotics", part 1, Humboldt University Berlin, October–December during Winter Semester 2012/13, 32 h, 10 participants. The only regular course of our evaluations.
4. Intensive Course "Cognitive Robotics", Vistula University Warsaw, February 25th–March 1st, 2013, 30 h, 30 participants.
5. Intensive Course "Cognitive Robotics", at University of Novi Sad, Serbia, March, 12–21, 2013, 32 h, 9 participants.
6. Intensive Course "Cognitive Robotics", at University of Rijeka, Croatia May, 21-29, 2013, 24 h, 18 participants.

At all events, lectures and practical exercises were mixed, and the exercises took about 30–50 % of time (see below).

1.10.1 Local Requirements for the Courses

As already discussed, RoboNewbie is intended for easy usage by beginners in Robotics. Therefore, the requirements for the users are as minimal as possible, while the framework gives maximal support.

Since most of the courses had only a short duration, organisational issues were important for the success. We have asked the local organizers to prepare the technical resources accordingly. In the following, we describe the requirements in more detail.

Participants: Users are expected to have some programming skills in Java, such that they are able to understand and modify the agent programs. The programs are already prepared for usage under Netbeans, therefore the participants should be familiar with such tools. Users should be able to download and instal programs from the web according to given instructions.

Some physical and mathematical background is needed to understand the theoretical and practical issues of Robotics. Related undergrade level is sufficient. Moreover, we plan the usage at Secondary Schools, but we could not test up to now.

Preferably, participants should work in teams (as useful for programming exercises in general). Each team might consist of 3–5 participants, preferably mixed by different skills of its members. It helps for a smooth course if there are no big differences between the teams (e.g. each team should have at least one of the good programmers of the course, good mathematicians etc.).

Technical Resources: The participants should have their own computers where they can install and use the programs. Participants need access to the computers during the courses as well as for their homework. Hence, laptops are preferable. They are sufficient to run all the programs. Alternatively, participants may use computers in a lab (which have to be prepared accordingly if students are not allowed to install their own software).

The list of needed installations is given on the RoboNewbie webpage (cf. Sect. 1.7). Instructions for installation and functionality tests are found there, too. If possible, students should get information before starting the course. They should be asked to install the programs by themselves and test if the programs can be started. If students can not be asked before, an on-site test by some responsible person should be performed. It helps to save time during the courses if on-site problems with hardware or software are solved before. Nevertheless, if computers are ready, installation of programs needs only short time and can be done at the beginning of a course.

Organisational Issues: A good schedule is necessary for smooth courses. This includes early information (as far as possible) of participants as described above. Then the lectures and exercises are mixed appropriately. After a short overview about Robotics, participants start their first exercises as given by the *Quick Start Tutorial* (also found on the RoboNewbie web page). Later, more explanations are given as far as the theoretical lectures proceed. Thus, theoretical introductions to sensors are connected to explanations of perceptor usage in RoboNewbie, introductions to motions are connected to the development of keyframes etc.

Competition: The courses end with a competition, which serves as a motivation for the participants. The successful participation at the competition can also be a substitute for an examination if students need some certificate.

The competition is announced at the beginning of the course, and it should be performed by the teams. This helps for the integration inside the teams from the very beginning. The number of competitors should be not more than 10 in order to make the contest not too long. This is also an argument to form teams if the number of participants is larger. The level of teams should be comparable for fairness reasons.

For the competition, the teams have to improve the `Agent_SimpleSoccer` program to get a better performance. It is up to them, what they want to improve. Actually, `Agent_SimpleSoccer` performs very poorly, it needs about 10 min to push the ball from the middle into the goal. It was designed this way just to motivate the participants for improvements, which can be achieved in many ways by changing the original program. The result should be a better performing agent which needs less time for scoring, e.g. by better walk, related kicks, better world model, better control, cooperation etc. At the competition, each team should give a short description of its efforts and expected results, at best just before its trial. This is also a possibility to check the engagement of each team member.

To make the competition a success (and a fun), it must be organized by strict and transparent rules. It should have a tight schedule to emphasize the aspects of sports. Therefore, each team has only one trial of only 3 min. The ranking of teams is determined by fastest scoring times. For teams who did not score, the ranking is given by minimal distances to the goal after the 3 min have elapsed.

Fig. 1.14 Impressions from the DAAD course in Ohrid, August 2012

To save time and to avoid compatibility problems, teams use their own computers in the competition. For fairness reasons, the teams have to stop working on their programs at the same time just before starting the competition (to avoid complaints, this should be controlled, e.g. by collecting the computers at a special place). The 3 min for each trial have to be measured by SimSpark (because computers might have different performances).

Moreover, each participant can be asked to provide a written report of his/her individual efforts. Contents of the reports should be the performed trials and its (positive or negative) results, and the contribution to final program of the team, respectively.

1.10.2 Evaluation and Results

We have asked the local organizers to prepare the courses according to our requirements as described in Sect. 1.10. It was not always possible, to install the programs before the beginning. But as far as the participants were prepared to do installations by themselves, they were ready to work after less than one hour following the steps described in the document "Installation".

The participants of the courses (except for the first short seminar) were asked to give feedback on a prepared form at the end of the course. They could evaluate different aspects of the course and the framework by numbers between 1 and 5, as given by Table 1.1.

As the evaluation shows, the exercises with the simulated robots were motivating and helpful, the participants wanted to have more time for exercises and especially for own experiments. As expected, the participants with less experience in Robotics gave higher marks related to motivation and help.

The usage of the framework was intuitive. Interestingly, the participants with more experience in Java programming gave significantly higher rankings.

The different times spent for homework were related to the special conditions of the courses. The courses in Warsaw and Novi Sad were held as compact courses during the semester, where students had to obey further obligations. In contrast, the course in Ohrid was a summer school, while the course in Berlin had a duration of two months.

The level of the exercises was considered as adequate, but for that the proportion of exercises was adapted by us accordingly.

As a unique observation, participants wanted to have more time for exercises than for lessons. This may have several reasons. The individual work load resulted in a bias for exercises: The participants had to fulfil given requirements, and many of them spent much time for preparing the final competition. This becomes obvious especially in Rijeka, where the amount of common exercises were extremely short. Furthermore, the lectures tried to give a broad overview about the actual state of art in Robotics. There was not enough time to exercise on all these topics. Hence, a reduction of topics in the lectures of such short courses should be

Table 1.1 Evaluation by participants of the courses

	Ohrid	Berl.	Wars.	NSad	Rijeka	
Duration	6	60	5	10	8	(days between start and finish)
Total time	20	32	30	32	24	(hours for lectures and exercises)
Participants						
Number of participants	26	10	33	9	18	
Java skills	2,74	3,2	2,84	3,83	3,67	(1 = none... 5 = many)
Robotics skills	1,83	2,8	1,58	1,33	1,30	(1 = none... 5 = many)
Exercises						
Time at course	11	12	15	10	5	(hours a 45 min.)
Time at home	9,81	23,3	5,4	4,8	8.7	(hours a 45 min.)
Quick start	4,09	3,4	4,16	4,5	3,93	(1 = not helpful... 5 = very helpful)
Motivation	4,57	3,4	3,74	4,67	4,13	(1 = boring... 5 = motivating)
Level	3,00	2,8	3,26	3,0	2,93	(1 = too easy... 5 = too difficult)
Participants want more time for						
Theory	2,96	3,0	2,94	3,16	1,53	(1 = less... 5 = more)
Advised exercises	3,78	3,8	4,11	4,0	4,27	(1 = less... 5 = more)
Own experiments	4,07	3,6	3,83	4,0	4,07	(1 = less... 5 = more)
Discussion	3,70	3,2	3,67	3,5	3,80	(1 = less... 5 = more)
Evaluation of RoboNewbie framework						
Structure	3,48	4,4	3,0	4,0	4,20	(1 = difficult... 5 = intuitive)
Usage of classes	3,95	4,6	3,61	4,33	4,00	(1 = difficult... 5 = intuitive)
Usage of documents	4,05	n.s.	3,79	3,67	4,40	(1 = not at all... 5 = very often)
Help by documents	4,18	4,8	4,39	4,5	4,47	(1 = not at all... 5 = very helpful)
Result for Winner of the competition						
Closest distance	3,5	5,3	5,7	1,5	0	(distance between ball and goal)
Time to score	–	–	–	–	57	(seconds if scored)

(continued)

Table 1.1 (continued)

	Ohrid	Berl.	Wars.	NSad	Rijeka
Did simulation help to understand real robots?					
Help by simulation	4,18	3,8	4,37	4,17	4,47	(1 = not at all… 5 = very helpful)
Which scenarios are interesting as a task for learning Robotics?					
Soccer	15	3	15	4	11
Rescue	11	1	2	3	5
Household	9	0	5	4	7
Other	3	1	2	1	1

Mean values according to the scale 1…5 are given. Hours consist of 45 min.

considered. On the other side, the e-learning material should be extended with related exercises.

The participants of the course in Berlin had more time (two months) for their studies and exercises. Students implemented more sophisticated methods and tried out changes of the framework itself (e.g. other interpolation methods for key-frames). This again came by no surprise, because they started with more knowledge in Robotics, while RoboNewbie is designed for beginners. As a consequence, the framework will be extended with more challenging experiments for advanced users.

When we will prepare related extensions we have to be aware about possible conflicts concerning didactics and motivation. Not all such experiments can be performed under the "minimalistic" conditions. For example, experiments with cooperative behavior should be based on sufficiently well performing skills, e.g. for walking or passing. The development of such skills (e.g. by physical or biological models, by Machine Learning etc.) is a challenging task. This would need time for development, and it opens further experimental fields in more advanced courses, too. But for experiments on cooperation, acceptable skills should be given to the participants. If participants have started with basic methods and have developed related keyframe motions before, then they may feel frustrated about their former "useless" efforts: Why to work hardly on exercises with keyframes, when better skills are available later. As an alternative, keyframes should be used in exercises for other skills, e.g. for dancing or for stand-up.

As another problem, the approximated correction of the coordinates (angles) according to head motions as described in Sect. 1.8.2 should be discussed further. It results in deviations when the head is facing downwards, e.g. when the robot looks for a nearby ball while preparing a kick. Moreover, the reported values change at different cycles while the facing direction changes. This could be misinterpreted as motion even if the ball does not move (actually the distance to the ball is not affected and remains constant, only the reported angles can vary).

The situation for a tilted camera is illustrated by Fig. 1.15. Actually, the head is rotated in the neck, which results also in a small translation of the camera which is placed in the center of the head. This translation is in fact small, and we can ignore it for our discussion. As can be seen, the horizontal angle and the vertical angle are both affected. Knowing α' and δ' from the vision perceptor data, we want to calculate α and δ as if the robot would face in forward direction. A precise correction would need calculations using the rotation by the head pitch angle ϕ.

Our approximation in the program LocalFieldView uses simply an offset by the turned head angles. This is correct if the head is only turned horizontally by some yaw-angle while the nick angle is zero. Differently, if the head is tilted, the situation is more complex as illustrated by the Fig. 1.15. Here, the correction by offsets gives only approximated results.

Better calculations could be provided for more experienced users which understand the geometry behind, or could be left as exercises for them. To provide

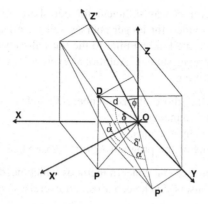

Fig. 1.15 The camera of the robot in point O is tilted by pitch angle ϕ while the yaw angle remains zero. The cartesian coordinate system X–Y–Z of facing forward is rotated by ϕ around the Y-axis to the new system X'–Y'–Z', with $Y' = Y$. The object D has cartesian coordinates (x, y, z) and (x', y, z'), respectively. It has polar coordinates (d, α, δ) in the original system. The distance is given by d, the horizontal angle α is the angle between the X-axis and the line OP, while the vertical angle δ is the angle between OD and OP. $P = (x, y, 0)$ is the projection of D to the X–Y-plane. The object has polar coordinates (d, α', δ') in the rotated system, and $P' = (x', y', 0)$ is the projection of D to the X'–Y'-plane. The distance is again given by d, the horicontal angle α' is the angle between the X'-axis and the line OP', while the vertical angle δ' is the angle between OD and OP'

such algorithms for inexperienced users would violate the desired transparency of the system.

As expected, the participants were able to use the framework after very short time of introduction, i.e. on the first day of the course. The documents *Quick Start Tutorial*, *HowToStart*, and *SimSpark* were used as first references. These documents are extended for the e-learning project.

In all courses, the final competitions were a source for fun, and many participants came with original ideas and solutions. Up to now, only one team from Rijeka could score in the given 3 min frame during the competitions, but some more could during the preparation. Of course, the competition is also affected by individual luck, because the scenario works not deterministically: Occasionally a robot falls down and the agent looses time. Nevertheless, the engagement of participants can be observed and evaluated even in such cases.

Even in case of few participants, the task should be performed as teamwork. In Novi Sad, the task was given as an individual work, which appeared to be difficult for some participants. Nevertheless, good results were obtained even there.

All in all, the evaluation has shown, that RoboNewbie is well suited for its intended usage as a framework for beginners. It allows first experiments with complex robots. More experienced users could follow their own ideas. Over all our courses, there were no substantial complaints by the users, and we did not find the necessity for substantial changes.

The amount of exercises was sufficient for our short courses. But additional exercises should be provided for longer courses and for e-learning. There should be more exercises of simple kind (similar to the ones already prepared), and more challenging ones for deeper studies. Additional exercises on the level of existing examples could concern e.g.

- Different behaviors like those of "Braitenberg-Vehicles",
- Balancing, knee bend, dancing, ...,
- Swarm behavior of simple agents,
- Further improvements of the program `Agent_SimpleSoccer`.

More advanced exercises can concern methods which need well educated users or even teams of such users. Such scenarios and exercises can be found e.g. for

- Motions by different paradigms (model based, biologically inspired motions, Machine Learning approaches),
- Advanced perception (world models, mapping, ...),
- Modelling other agents,
- Coordination strategies,
- Competitive soccer players (e.g. for RoboCup competitions).

The soccer scenario as a well known scenario was well suited for our introductory courses because it needs the integration of motions, perception and control skills. We did not address all challenges (dynamically changing situation, real time behavior, cooperation with team mates, controversial agents, ...) and hence there is still room for further challenging tasks. Nevertheless, it would be useful to have other environments like rescue robots or home service robots, but the efforts for the implementation are behind our actual personal resources.

1.11 Conclusion

In contrast to other experiments in Robotics, the RoboNewbie experiments can be performed without special hardware. It simply needs a computer for the simulation of complex robotic scenarios. Actually, the scenario is more complex than the scenarios provided by many available hardware equipments. It is easy to understand and to use after a short introduction. No special knowledge (except basic programming in Java) is required to start with own experiments, and while the users acquire more knowledge, they can work on more challenging tasks.

The practical evaluations have confirmed our expectations on the RoboNewbie project. Beginners in Robotics were able to use the framework after short introductions. They were able to program own methods in parallel to the theoretical concepts and methods provided by classes. Participants have attested the usefulness of own experiences.

The "minimalistic approach" is useful especially for short courses and for introductions to longer courses. Later on, more sophisticated methods will be useful for higher level integrative tasks. It is impossible to let students implement all desirable algorithms in the limited time of a course. Joint activities of robots, for example, depend heavily on the available bodily skills and on the capabilities for interaction and coordination. Hence related skills for usage in the courses should be prepared before. It will be done while the project is developed further for e-learning purposes.

Next plans concern the evaluations of the RoboNewbie framework in Secondary Schools, and the integration into an e-Learning course on Robotics. The web page with its documents is an already tested starting point for the related exercises. We also hope, that RoboNewbie can serve as an entrance to the 3D simulation league of RoboCup, such that it comes back to its roots.

References

1. Aldebaran: Manufacturer of the robot Nao. http://www.aldebaran-robotics.com/en/. Visited at 27.3.2013
2. Aldebaran Robotics. NAO Software 1.12 Documentation. http://www.aldebaran-robotics.com. Visited at 7.2.2013
3. Braitenberg, V.: Vehicles: Experiments in Synthetic Psychology. MIT Press, Cambridge (1984)
4. Darpa Challenges: Urban Challenge. http://archive.darpa.mil/grandchallenge/, Robotics Challenge. http://www.theroboticschallenge.org/. Visited at 27.3.2013
5. Koenig, N., Howard, A.: Gazebo - 3D multiple robot simulator with dynamics. http://playerstage.sourceforge.net/gazebo/gazebo.html (2004). Visited at 7.2.2013
6. Lego Mindstorms. http://mindstorms.lego.com. Visited at 23.5.2013
7. magmaOffenburg Homepage. http://robocup.hsoffenburg.de/html/index.htm. Visited at 23.5.2013
8. magmaOffenburg Code Release. http://robocup.hs-offenburg.de/downloads/magma3D-2011Release.tar.gz (2011). Visited at 23.5.2013
9. Microsoft Robotics Developer Studio (MDRS). http://www.microsoft.com/robotics/. Visited at 23.5.2013
10. Padir, T., Chernova, S. (eds.): Special issue on robotics education. IEEE Trans. Educ. **56**(1). http://ieeexplore.ieee.org/xpl/tocresult.jsp?isnumber=6423944 (2013). Visited at 23.5.2013
11. Pfeifer, R., Bongard, J.: How the Body Shapes the Way We Think: A New View of Intelligence. MIT Press (2006)
12. RoboCup: official web page. http://www.robocup.org/. Visited at 27.3.2013 (The annual proceedings since 1997 are published as "RoboCup [year]: Robot Soccer World Cup [number]" in the Subseries *Lecture Notes in Artificial Intelligence* of the Series *Lecture Notes in Computer Science*, Springer)
13. RoboNewbie. http://www.naoteamhumboldt.de/projects/robonewbie/. Visited at 27.3.2013 (A detailed description is given in: Monika Domanska. RoboNewbie. Robotersimulation für Lehre und Forschung. Submitted as Diploma thesis, Humboldt University Berlin, 2013 (in German))
14. Robotis: Manufacturer of the Bioloid toolkit. http://www.robotis.com/xe/bioloid_en. Visited at 2.5.2013
15. ROS (Robot Operating System). http://www.ros.org/. Visited at 2.5.2013

16. SimSpark RCSS Wiki (Documentation of the Simulator). http://simspark.sourceforge.net/Wiki. Visited at 27.3.2013
17. Simulator of the 2D Soccer Simulation League. http://sourceforge.net/apps/mediawiki/sserver/. Visited at 23.5.2013
18. Smith, R.: Open dynamic engine user guide. http://www.ode.org (2006). Visited at 27.3.2013
19. Stone, P.: RoboCup as an introduction to CS research. In: Polani, D., Browning, B., Bonarini, A., Yoshida, K. (eds.) RoboCup-2003: Robot Soccer World Cup VII, Lecture Notes in Artificial Intelligence, pp. 284–295. Springer, Berlin (2004). The material of the course "Autonomous Multiagent Systems" at the University of Texas (2012) can be found at http://www.theroboticschallenge.org/. Visited at 23.5.2013
20. USARSim (Unified System for Automation and Robot Simulation). http://sourceforge.net/apps/mediawiki/usarsim/. Visited at 23.5.2013
21. Walter, W.G.: An imitation of life. Sci. Am. **182**(5), 42–45 (1950)
22. Walter, W.G.: The Living Brain. W. W. Norton, New York (1963)
23. Webots by Cyberbotics. http://www.cyberbotics.com/. Visited at 2.5.2013

Chapter 2
Designing Intelligent Agent in Multilevel Game-Based Modules for E-Learning Computer Science Course

Kristijan Kuk, Ivan Milentijević, Dejan Rančić and Petar Spalević

Abstract Nowadays, game-based learning environments are very common environments for studying major scientific fields such as mathematics, computer science, electronics and electrical engineering. This chapter presents a game-based modules system called the game-based modules (GBMs). It combines the characteristics of computer game elements with the existing interactive multimedia environments for learning mathematics, physics and electronics. This module presents a new type of game-learning environment for teaching units of Computer Science courses. Bearing in mind that the GBMs includes interactive tasks as a form of a multi-level approach to problem solving, we have also shown an approach to evaluating student's knowledge necessary for upgrading him/her to the higher level of learning. To assess a student's knowledge level needed for the next game level in the GBMs, we have developed an intelligent agent. This illustrates how intelligent agents and fuzzy logic can help increase the quality and quantity of the most important element of e-learning and that is making a decision. The results of student's knowledge diagnosis by means of agent within the GBMs e-learning system demonstrate the possibility of applying the presented agent model in various game-based learning systems for the determination of the knowledge level performance. On the basis of the data obtained through the exams, as well as through the use of statistical reasoning methods, we have shown the efficiency of the GBMs in the learning process.

K. Kuk (✉)
School of Electrical Engineering and Computer Science Applied Studies, Department for New Computer Technologies, University of Belgrade, 11000 Belgrade, Serbia
e-mail: kukkristijan@gmail.com

I. Milentijević · D. Rančić
Faculty of Electronic Engineering, Department for Computer Science,
University of Niš, 1800 Niš, Serbia

P. Spalević
Faculty of Technical Sciences, Department for Electrical and Computing Engineering,
University of Pristina in Kosovska Mitrovica, 38220 Kosovska Mitrovica, Kosovo

M. Ivanović and L. C. Jain (eds.), *E-Learning Paradigms and Applications*,
Studies in Computational Intelligence 528, DOI: 10.1007/978-3-642-41965-2_2,
© Springer-Verlag Berlin Heidelberg 2014

Keywords E-learning system · Pedagogical agent · Game-based modules · Intelligent agent model

2.1 Introduction

Game-based learning promises to be a successful approach to teaching computer science courses. Educational games and interactive simulations can enable a student to acquire knowledge in a specific field by playing a game successfully. To a great extent educational games can be referred to as computer science disciplines. Authors Shabalina et al. [1] have implemented the educational games concept in the learning game for C# programming language. Their system is based on the common game engine architecture, but it has been extended to the use in educational games and it consists of two high-level subsystems: a game engine and a learning engine. Computer Programming course is found to be difficult and boring. Thus, learning through games seems to develop students' motivation for the subject. Authors Roslina et al. [2] describe the perceptions of students at the Malaysian university (UTM) about using educational games for the purpose of self-learning within introductory programming courses. Virtual learning environments support teaching and learning in an educational context, offering the functionality to manage the presentation, administration and assessment of coursework activities. Callaghan et al. [3] demonstrate how immersive virtual worlds can be used for a game-based strategy for the purpose of teaching electronics and electrical engineering by using a collaborative team-based competitive format. Studies conducted in a Greek High School within the Computer Science course investigated potential gender differences with respect to the game-based learning effectiveness, as well as with respect to the motivational appeal of a computer game to learning computer memory concepts [4].

A learner model, also known as a student model, refers to the model constructed from observing the interaction between a learner and a learning system or instructional environment. A student model must contain the following important information about the user: domain knowledge, learning performance, interests, preferences, goals, tasks, background, personal traits (learning styles, aptitudes...), environment (work context) and other useful features [5]. Domain-specific information is organized into a knowledge model. The knowledge model has many elements (concept, topic, subject...) which students need to learn. In this chapter we have described the learning strategy for computer science curricula as a possible model for college students.

The strategy is founded on the research conducted in the field of teaching methodology on the one hand, and game-based learning features on the other. The learning environment should be customized to the individual learner's learning styles and educational needs with the quality of the learning experience continually validated and evaluated. Software agents can be used to support instructors and domain experts with both course design and delivery. They can also support individual learners by personalizing course materials based on learning objectives, learners' characteristics, and learners' prior knowledge, and facilitating learners' interaction. Agent technology, a combination of artificial intelligence and software engineering, represents an exciting new means of analyzing, designing and building complex software systems [6]. Agent-based systems have been successfully used in many areas such as information collection/filtering, personal assistants, network management, electronic commerce, intelligent manufacturing, health care, entertainment, etc. [7, 8]. An agent works towards its goals. The agent's goal model ensures the agent will do the right thing at the right time. Agent technology has existed for a long time, but there are few researches combined with educational values or educational technology. Actually, personalized or adaptive game-based learning has become very popular in the game-based learning and game-based testing. This chapter illustrates the easy integration of agent technology into game-based learning system and a case study based on it.

Section 2.2 discusses in more detail the related work we consider most relevant. Section 2.3 presents some pedagogical elements of the student model realized through the Net-Generation students and Game-Based Learning (GBL); Sect. 2.4 includes the use of game-based modules—the GBMs as a new teaching approach to teaching units in the Computer Science courses. This section shows two types of multi-level GBMs: (a) module for the "Unary logical operations" teaching unit and (b) module for the "Z-buffer" teaching unit. Section 2.5 describes the proposed intelligent agent model for the assessment of the knowledge level for the game-based learning system, the results of which are estimated by an equation with variable coefficients. Finally, Sect. 2.4 illustrates experimental results of estimated values obtained through the agent model versus the results obtained through classical tests and the efficiency of the GBMs as a supporting tool for the preparation of the exam questions.

2.2 Related Work

Recent research into game-based learning has identified adaptability as an area that requires further attention. Adaptability is required in game-based learning simply because each person has a different way of learning in different learning environments—one size does not fit all [9, 10]. Thus, learning may be related to and influenced by the player's preferences and customization of elements within the learning environment. The IMS Learning Design (IMS-LD) is a specification for creating Units of Learning (UoLs) which express a certain pedagogical model or strategy (e.g., adaptive learning with games). In this sense the MS-LD can be complemented with off-the-shelf components and resources integrated in Units of Learning (UoLs). Author Koper [11] in the project named <e-Adventure> shows how an adaptive IMS-LD UoL can be modeled and integrated within an external resource such as educational game. The main goal of the project was to apply a documental approach to the development of educational adventure video games (often also referred to as point-and-click adventure games or conversational games). However, the question is what sort of adventure games should be included in the curriculum and what degree of complexity should they offer. The authors Jeetinder and Jayanthi [12] responded to this question by offering a framework for creating individual simulations (modules) and organizing them in the form of multiple levels of a game. For educational games to be effectively integrated into the curriculum, they propose to concentrate on developing simple and small games and make explicit the relation the game bears to the lessons. This is better than creating very complex games with which the student is left to figure out the relation. As a result, this chapter shows how to create simple modules in the GBMs system as independent and interrelated concepts in the field of Computer Science.

Adaptive game-based learning is a fundamental issue for the next generation of educational games where progress is controlled in accordance with the learners' behavior. The major aim of an adaptive game-based learning system is to support and encourage the learners considering their needs, strengths and weaknesses. In their paper Weng et al. [13], proposed the framework of a personalized Quiz-MASter assessment game and the flow of a personalized quiz game. With the services provided by an intelligent agent, a user can play the personalized assessment game by means of the flow of a personalized quiz game. Actually, the personalized or adaptive game-based learning has become very popular in the game-based learning and game-based testing. In the GBMs system proposed in this chapter an adaptive system approach to users is being realized through a pedagogical agent. Within the system the agent presents the modules on the basis of the estimated knowledge of every student.

Pedagogical agents are embodied software agents that have emerged as a promising vehicle for promoting effective learning. They provide customized problem-solving experiences and advice that are precisely tailored to individual learners in specific contexts. However, Conati and her colleagues [14] believe that the pedagogical agent could strike a better balance between learning and engagement if, in addition to the student's knowledge, it could have access to a student's affective reactions to the game. Bayesian techniques must be used when the agent's decisions about the states of the game world are uncertain, because a Bayesian network is often used to model the agent's uncertainty about its opponents. A project by Lester [15] at North Carolina State University focuses on the development of a full suite of Bayesian pedagogical agent technologies for inquiry-based science learning. It will provide a comprehensive account of the cognitive processes and results of interacting with Bayesian pedagogical agents.

In regards to Bayesian techniques, the implemented pedagogical agent in our e-learning system uses a simple fuzzy logic deduction—"if-then". Intelligent agents in combination with fuzzy logic can help increase the quality and amount of interaction in a computer game. The storyline often demands precise control over certain creature's properties, but autonomous agents may exhibit undesirable emergent behaviors due to the absence of centralized planning and control. In order to achieve this, our agent uses the students' knowledge assessment module in the section about pedagogical module. The mentioned pedagogical module intended for calculating the current learners' level of knowledge in the GBMs system uses a simple mathematical formula, unlike some other agent-based game design technologies which often use the given BDI.net agent framework [16].

2.3 Method

2.3.1 Problem Statement

Upon analyzing the final exam results during the examination periods in the Computer Graphics course at the School of Electrical Engineering and Computer Science Applied Studies in Belgrade, we came to the conclusion that the results of one group of questions were much more different than the others. Performing an analysis we discovered that those were the questions referring to the field of hidden surface techniques and especially to the Z-buffer algorithm. Although students had the same learning materials for the purpose of the exam preparations in all fields, the difference discovered in this teaching unit showed that this field was quite complex and abstract for students. Therefore, it was necessary to take some steps in order to improve the approach to learning regarding this teaching unit, as well as to improve the final exam results. Of all the algorithms for finding visible surface, the Z-buffer algorithm is perhaps the simplest and therefore most frequently used. Starting from the facts that this teaching unit is simple and that, in spite of that, the students show worse results in this field than in any other, we began searching for the ways how to offer our students the teaching material which would be more student-friendly.

During laboratory exercises in the Computer Architecture and Organization course students brush up and improve their knowledge acquired in lectures through concrete tasks and under the supervision and help of assistant lecturers. Taking exercises without previously acquiring the basic terms in the field being taught in the lectures directly results in difficulties with individual realization of the laboratory exercises. The possibility of visual representation of the task solving method for rehearsing materials in this course enabled their implementation in the form of an interesting game. For the purpose of motivating students to get ready for laboratory exercises and get them more active in individual task solving during the exercises, the educational game ArhiCOMP has been created. This educational game contains interactive tasks [17, 18] implemented in the graphic environment, which directly associates it with the field of application use. The game has been designed to help the students learn basic terms referring to unary logical operations and to apply them practically through solving the given examples by randomly generated content of virtual registers, which are an integral part of the arithmetic-logical unit of computer systems.

2.3.2 Characteristics of Students

Nowadays, elementary and secondary school pupils, as well as university students, belong to the generation born in the Internet age. Modern psychologists, sociologists and pedagogues refer to them as the Net-generation. For the above-mentioned reasons, in this work we tried to make it easier for students to learn abstract educational materials and to improve their score in the final exam by using any type of interactive multimedia applications. The Net-generation students prefer learning by being told what to do [19]. They learn successfully through discoveries—both individually and with their age-mates. They learn by doing, and not by reading instructions from the manual or by listening to lectures. The Net-generation is more comfortable in the environment abound in pictures than in text. The researchers' report that the Net-generation students will refuse to read a large quantity of text, regardless of the length of the task or instruction. Rather they merely think or speak about activities they like to perform them. Each student has a different learning style. Some students learn best by visual learning exercises. These (visual) learners benefit from a variety of ocular stimulation. One example would be the use of color. Others do best by performing intellectual activities—problem solving and reasoning. Intellectual learners like to engage in activities such as solving problems, analyzing experiences, doing strategic planning, generating creative ideas, etc. [20].

In order to make it easier for students to learn and acquire knowledge in the computer science courses that cannot be seen with the naked eye or observed on the basis of pictures, we used the characteristics of two types of students: intellectual and visual. Bearing in mind the fact that the intellectual-type learners are fonder of educational material in the form of simulations and interactive tasks, we created our environment for learning the Z-buffer algorithm as an environment in which the task solving is performed as a simulation of the operation of the mentioned algorithm. On the other hand, taking into consideration the fact that visual-type learners remember graphically presented information more easily, the environment was enhanced with various colors and shapes for concepts as integral parts of the algorithm. We reduced the definition of registers content used by the algorithm to decide upon the color, and the register itself was presented as a series of squares, where each square stands for 1 bit.

It is considered that educational games improve learning due to a better learning method that meets the needs and habits of the Net-generation students. To provide this kind of support for learning is at the same time extremely important and represents an extreme challenge. Creating special learning systems that contain educational games represents a challenge since it demands a careful analysis of stimulating learning and maintenance of positive motivation.

2.3.3 Simulations and Games

A memo-technical rule says that in order to memorize a term or a definition more easily, they should be made interesting. Motivation is also increased when we do something amusing. In this regard, one of the significant techniques is visual representation. If a piece of information can be represented by a visual picture or animation, then it should be represented in that way. Simulations and games, as highly interactive multimedia applications, can increase students' motivation for science learning, deepen their understanding of important science concepts, improve their science process skills, and advance other important learning goals. Through solving the tasks in the form of interactions the students are required to recognize the simulation of the operation of an algorithm or process, but they are not afraid of giving incorrect answers or performing wrong actions.

After performing an analysis of the existing Internet simulations and other interactive multimedia applications used in the teaching process, we came to the conclusion that it is necessary to introduce some of those contemporary teaching resources to the course in Computer Graphics, so that students could learn the planned material in the best possible way. However, since the students to whom such a type of education material is to be presented do not belong to the generation for which the existing applications and modules had mostly been made, we had to introduce additional pedagogical elements to those applications to make them more acceptable for the students. On the one hand, these applications certainly have to be amusing, while, on the other hand, they must have an educational character.

Regarding the pedagogical elements supported by the games-based learning, Norman [21] identifies seven basic requirements for a learning environment: (1) to provide a high intensity of interaction and feedback, (2) to have specific goals and established procedures, (3) to motivate, (4) to provide a continual feeling of challenge that is neither so difficult as to create a sense of hopelessness and frustration, nor so easy as to produce boredom, (5) to provide a sense of direct engagement, producing the feeling of directly experiencing the environment, directly working on the task, (6) to provide appropriate tools that fit the user and tasks so well that they aid and do not distract, and (7) to avoid distractions and disruptions that intervene and destroy the subjective experience. By adding certain inherent engaging elements, suggested by Prensky [22], we created the most acceptable learning model for our type of students. As a combination of visual and intellectual types of learning, the game-based learning environment with its characteristic given in Fig. 2.1 will improve their psychological capability of photographic memory and help them to create mental images of concepts being learnt.

Stimulated by modern technologies, teachers use computer games and simulations more and more frequently in order to motivate students. Although video games can potentially be beneficial for learning, they must be aligned to specific curriculum content to achieve solid gains in learning. The three factors influencing the possible choice of the existing interactive multimedia applications for learning the subject were: content, age and learning styles.

Fig. 2.1 Characteristics of GBL environment that are used to help psychological student's phenomena

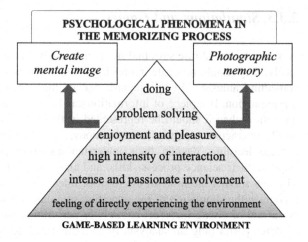

2.4 Game-Based Modules

The analysis of the existing multimedia interactive environment for learning in the field of education shows that there are three types of applications used effectively for teaching mathematics and computer engineering to students. Those types of existing educational applications are:

- Gizmos,
- IMMEX,
- Interactive tasks.

1. GIZMOS: ExploreLearning's Gizmos are an effective and engaging way to move students to inquiry-based science and math. The students get to see real simulations that they read about in textbooks. It gives real life problem-solving a whole new meaning! Interactive simulation that makes key concepts easier to understand and fun to learn [23]. World's largest and most advanced online repository of math and science simulations for grades 3–12. Gizmos are online simulations that are powerful teaching and learning tools to engage students. Their easy-to-use format makes them practical and effective. Students manipulate key variables, generate and test hypotheses, and engage in mathematical inquiry. Gizmos supplement and enhance your instruction with powerful visualizations of math concepts. Gizmos are an effective and engaging way to move students to inquiry-based science and math [24]. Gizmos move students to use higher-level thinking skills.

2. IMMEX: Interactive Multimedia Exercises (IMMEX) is an online library of science simulations that incorporate assessment of students' problem-solving performance, progress, and retention. Each problem set presents authentic real-world situations that require complex thinking. Originally created for use in medical school, IMMEX has been used to develop and assess science problem

solving among middle, high school, and undergraduate science students as well as medical students. Students navigate a hierarchy of menus and submenus to select different pieces of information which are each available at a cost. This form of task structure and subtask boundaries would be expected to elicit different levels of cognitive engagement as students explore and seek out relevant information [25]. The use of IMMEX software has been scientifically shown to have significant positive effects on students' understanding of science content as well as the process of scientific investigation [26]. By integrating our multimedia simulations into a unique web-based learning platform for modeling strategic thinking and problem solving, IMMEX is able to help teachers.

3. INTERACTIVE TASKS: Through an adaptive approach, with visual indication of the course of performing the task, students are enabled to learn the procedures for solving tasks from various fields and of various complexity levels. The very applications are created in the well-known software package for creation of multimedia content Adobe Flash CS4. Students' interactivity with this type of applications was achieved with the use of graphic symbols (different types of buttons) and specific colors. Realized interactive tasks [27] differ in: (1) The way of giving answers, (2) Visual guidance during the task execution, (3) The level of complexity. Interactive tasks will not allow advancing to the next level of solving unless the student previously fills or selects interactive fields in the previous level accurately. From the pedagogical aspect, these applications stimulate students to make conclusions on the accuracy of achieved steps in solving of the entire task on their own. In this interesting way, through self-revealing students acquire the needed knowledge quantity and thus carry out the learning process without tutor's intervention or previous knowledge, through random guessing.

By combining characteristics of these three types of educational applications, we have created a model learning environment (Fig. 2.2) that will be acceptable for implementation of teaching units in the course Computer Science, on one side, and for previously analyzed type of students, on the other side. In order to increase the engagement and interest of students for this type of teaching material, we included game characteristics in this environment. Thus the game concept should be based on two components: (a) learners must get the course information through its interpretation in the game world; (b) learners must see the result of this algorithm in a game context. Also, besides placing a game interface into learning environment we have also applied basic game elements, such as: result, time and difficulty levels. These new modules, which include game elements, represent research multimedia learning applications and are intended for Computer engineering students. These new learning environments, which include game elements, represent research multimedia learning applications and are intended for Net-generation students, we named game-based modules (GBMs).

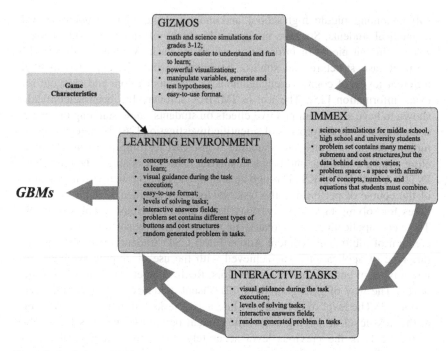

Fig. 2.2 Evolution of GBMs

2.4.1 Implementation of GBMs

The basic terms that explain the principle of operations functioning are reached by selecting the Help option in modules. When a student starts learning with the use of the game or faces a difficulty during solving a task generated by the application, Help serves to accelerate finding the right solution. This means that formulation of definitions and theorems within Help is the key moment in designing the entire application. The purpose of learning through the game is to enable students to learn the rules and check them in practice on the example of all unary operations. Multiple repetition of tasks with performing the same operation increases the probability of learning characteristics and use of particular operation. Quality evaluation whether an operation is acquired or not is performed through visual indication of the number of successful and unsuccessful tasks (score) with the same operation, and comparison with preset criteria. The model of the e-learning system GBMs is shown in Fig. 2.3.

2.4.1.1 Z-Buffer Module

The possibility of visual representation of the task solving method for rehearsing material in the course Computer Graphics enabled their implementation in the

Fig. 2.3 The model of the e-
learning system GBMs

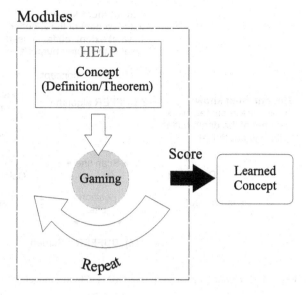

form of game-based module. The game has been designed to help students learn basic terms referring to the principle of the Z-buffer algorithm operations and practically apply them, through solving given examples with randomly generated content of buffer registers. This GBMs contains interactive tasks implemented in the graphic environment, which directly associates with the field of the application use.

By using advantages of the concept map technique, on the course in Computer Graphics we created a concept map for the teaching unit Z-buffer, with the aim to reduce the items presented in this unit to main concepts and to connect them in the simplest possible way. Having in mind the terms students should learn in order to successfully acquire knowledge about the Z-buffer algorithm functioning and thus have the needed knowledge for the exam, prior to creation of the conceptual map we marked this as the needed knowledge (Fig. 2.4).

Starting from the theoretical basis of this algorithm functioning, its integral parts (sub-concepts), without which the algorithm cannot function, are the concepts it contains—Z-buffer memory, Screen buffer memory and Scanning line. Every memory type (including the two mentioned) consists of a series of buffer registers containing a series of bits. This type of algorithm uses a 16-bit buffer, and the content of bits depends on the functioning principle of two types of tests, which are key sub-concepts of this teaching unit. In case that a Depth test is used for determining the depth of the observed polygons in relation to the observer, the test results are placed in the buffer register bits. On the other hand, results of the Color tests are also put into the buffer register as a series of 16 various bit values. As shown on the conceptual map, the basis of this algorithm functioning is determining the bit value content (1 or 0). Determining the bit that is active and whose content is to be filled determines the scanning line position in the algorithm.

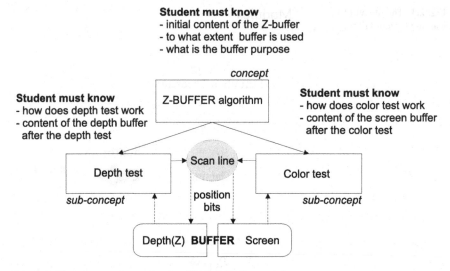

Fig. 2.4 The conceptual map for the GBMs Z-buffer module

Having in mind the concepts a student should learn, the student should connect these concepts through GBMs and successfully solve the given tasks. The tasks are given as a part of the game, and in the role of players students are motivated to solve them in order to advance in the game. Since two main concepts are given in the concept map (the depth and color test concepts), which are to be learned by students through this module, we presented them as two levels with different solving difficulty. Skill-based learning as a part of the micro game cycle completely fits into levels within the game. Many learning characteristics can occur during the game when the player attempts to go from one level to another. Adding time as a game element that contributes to the player's better ranking in the game also results in an increase of knowledge and improvement of the skills acquired through easier levels.

With the use of techniques for the first class innovative testing select/recognize [28], we came to the idea that answers in the GBMs can be given as a series of image fields on which a student should click. Since the buffer we use in the Z-buffer algorithm uses 16 bits, the task of this module is to determine the value of each bit, i.e. contents of the registry in various situations presented in the interactive task. The correct answer to fill the content of one bit is one of the proposed answers presented to students in the form of squares to be selected [29]. These squares, i.e. offered answers, are presented in two ways. In the level 1 of the module answers are offered in the form of a 13-square column (type 1). When certain square is selected in the scanning line (which shows the current active field in the task), the square falls down and fills the content of the active bit in the buffer. Answers in the level 2 are also offered as an array, but in the form of a five-square row (type 2). This row with offered answers is not constantly visible, but is shown only when the given bit is selected as a sub-menu in the menu list. The task at this game level is to determine the color in the screen buffer by determining the resulting color in each individual bit. The resulting color should be the result of overlapping of two or three pixel colors as parts of overlapping polygons in the given scanning line.

2.4.1.2 Module ArhiCOMP

The teaching unit "Unary logical operations" is aimed at teaching students about the way of performing logical operations at the level of registers in the computer system, through comparing the register binary contents before and after performing of the given operation. When it comes to operations for moving to the left or right, what is illustrated is the way of hardware implementation of the arithmetic operations division or multiplication with a degree of number 2. If students want to know how to apply an unary logical operation, i.e. if they want to know the register contents after the applied logical operation, they have to know basic rules of the binary digit system and rules referring to unary logical operations, such as the rule for logical shift to the right. Acquisition of basic terms is facilitated with the use of appropriate graphic representations, which presents the contents of the accumulator register in the arithmetic-logical unit before and after the shift operation. The arrows show the moving direction and new positions of bit in the register, as well as the contents of Carry flag in the condition register. Good knowledge of the set of rules from the field of unary logical operations is a precondition for future successful acquisition of knowledge in other fields within the course Computer Architecture and Organization or related courses in higher years of studies.

Starting from the fact that a well designed visual environment can attract the attention of students and motivate them to spend more time solving tasks within the educational game, special attention was paid to selection of the background, which in this case consists of various forms of binary statements presented in bright colors with effects of brightness, transparency and reflection, characteristic for new "fancy" technologies. To make working in the application more interesting, components in the game are not fixed but can move on the screen independently, which gives students comfort in the process of task solving. Moving and overlapping of components such as registers enables easier defining of their contents when a complex operation is applied. The task of the game is to determine the contents of Register 2 (which represents Register 1 immediately after the selected operation) in relation to contents of Register 1, which are randomly generated and appear after selection of the unary operation, together with the task text. When the student selects a bit in Register 2, the falling menu enables entering of the particular bit in the virtual register through selection of one of the options 0 or 1. Text of the task constantly changes on the basis of randomly selected values presented in the very text. Attempts to solve the task with the same text once again are reduced to a minimum. Observed from the pedagogical side, the repeated task solving prevents mechanical solving, but stimulates students to show the real level of acquired

Fig. 2.5 The help option in
the GBMs module ArhiCOM

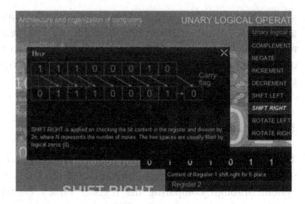

knowledge through previous problem solving. The basic terms that explain the
principle of logical operation functioning are reached by selecting the Help option
in the game (Fig. 2.5). When a student starts learning with the use of the game or
faces a difficulty during solving a task generated by the application, Help serves to
accelerate finding the right solution. This means that formulation of definitions and
theorems within Help is the key moment in designing the entire application.

2.5 Student Modelling in E-Learning System GBMs

In tutoring systems, students can learn new concepts and recognize the relationships between the previously learned and new concepts. This knowledge is
represented as a conceptual map in the GBMs system. Hwang [30] introduced the
"Concept Effect Graphs" where the subject materials can be viewed as a tree
diagram comprising chapters, sections, sub-sections and key concepts to be
learned. In order to obtain maximum benefits from the material, the teacher must
perform a very difficult task referring to the use of conceptual maps in the realization of the teaching units. In the process of designing and creating the teaching
units a lecturer may find concept maps very useful. Global "macro maps" can also
be made, showing the main ideas we want to present during the entire course, or
specific "micro maps", showing the structure of knowledge for specific fields. The
concept maps are graphic tools for organizing and presenting the knowledge base.

The conceptual map approach offers an overall cognition of the subject contents. The diagnosis process can be easily implemented through this approach. If a
student fails to learn the concept "sub-concept 1/level 1", it is possible that the
student did not learn the concept "sub-concept 2/level 2". Therefore, the system
suggests that the student needs to study the sub-concept 2 again. For this reason,
the GBMs knowledge base of the must show the relationships between concepts.
To do this, a conceptual map-based notation is proposed. Let us suppose that C_i
and C_j are two concepts and if the concept C_i is a prerequisite for the concept C_j
then the concept-effect relationship $C_i \rightarrow C_j$ exists.

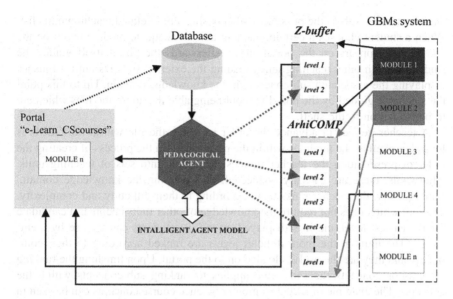

Fig. 2.6 Pedagogical agent in GBMs

The repository of knowledge stored in the knowledge base of the GBMs is structured as follows: a set of sub-concepts that represent atomic content teaching unit parts—concepts, a concept map that describes the interdependence between the sub-concepts. Each sub-concept is composed as a module including interactive tasks as game levels. Bearing in mind that the GBMs have interactive tasks implemented as multi-level problem solving, we have also shown the model for evaluating the knowledge necessary for upgrading to the next level, as shown in Fig. 2.6. To assess a student's knowledge level needed for the next game level in the GBMs, we have developed an intelligent agent model for the knowledge level assessment. By means of the intelligent agent model we have managed to assess a student's knowledge at the current game level, which has provided us with the basis upon which it is possible to decide if a student should move to the next level of learning or if he should stay at the same level.

2.5.1 Pedagogical Agent and System

The student is introduced to the set of all realized GBMs through the Internet portal "e-Learn_CScourses". The GBMs represent separate concepts that are being loaded on to the portal. Every module contains two or more separate files loaded within the module. The files represent the external *swf* files, while each one of them comprises a sub-concept within the module. Upon a student's registration and logging on to the portal, the Internet browser shows the Flash-supported environment as well as the possibility of choosing a pre-defined teaching units list. When a student chooses a certain teaching unit an adequate module based on the chosen unit appears on the portal shell. Afterwards the portal itself guides the student through other teaching units, loading the external *swf* files until it finishes displaying the associated sub-concepts in the knowledge domain. Up to this point the student is guided by the portal without being able to choose the files which are to be loaded on to the shell.

A teacher defines in advance the order in which the file within the GBMs is loaded, as well as the modules within the portal shell. In the process of creating the order of presentation the teacher divides the course into sets of teaching units. Afterwards, these are causally connected, thus creating the knowledge domain. Subsequently, the teacher ranks them according to their difficulty and complexity. The unit which does not require the knowledge of other units becomes a candidate for the easiest level and it is displayed on the list students can choose by themselves. The units at the more difficult levels are ranked according to their complexity and they are themselves loaded on to the portal. Upon finishing the ranking of units or concepts the same rule applies to ranking sub-concepts within the concepts. The example of sorted Computer Science course concepts can be seen in Table 2.1.

Since in the process of learning a student needs to show adequate knowledge about a specific concept or sub-concept, the next step in the assessment of causal relationships relating to the domain is to determine the student's knowledge necessary for the transition to the next concept, i.e. to the module itself or some of the game levels within the module. On the portal this task is performed by a pedagogical agent. The pedagogical agent assesses a student's current level of knowledge in the system and on the basis of this assessment it loads external databases into the system. Within the system the agent is never revealed to the student nor does the student know that the agent is being used by the system.

Table 2.1 Classification of certain concepts within the computer science course

Concepts	Sub-concepts	Complexity	Precondition
Data representation		1	No
Computer organization		1	No
Binary arithmetic	Add	1	No
	Subtract		
	Multiply		
	Divide		
Operation on bits	Unary logical operations	2	Binary arithmetic
	Binary logical operators		
Data manipulation		2	Computer organization
Machine language programming		3	Data manipulation
			Computer organization
Display systems		1	No
Z-buffer algorithm	Depth test	2	Display systems
	Color test		

Related modules are loaded independently and students do not decide upon the loading order. It is done by the agent on the basis of a conceptual map given by the teacher as well as on the basis of the knowledge assessment model.

While using the loaded module, on the basis of a student's logging on to the portal the data on his interaction with the modules are stored in the system database. Upon finishing the module game level the pedagogical agent receives the information about a student's interaction with the system and then by means of a simple "if-than" rule reaches the decision to let the student move to the next game level in the same or different module (thus loading another module on to the shell). The fuzzy rules are interpreted as:

if $(P(X) > \, = necessary\_KNOWLEDGE)$ *then* $(SHELL = next\_LEVEL)$

else $(SHELL = current\_LEVEL)$

If the agent decides that a student does not possess sufficient knowledge to move to the next level (sub-concept), the agent does not load the next level (sub-concept) within the same module (concept), but forces the student to return to the same level. The teacher determines the level of knowledge—P(X)—that the student needs to master on the basis of the analysis of the connection between the concepts in the knowledge domain. However, the question is how to calculate the current student's knowledge in the GBMs system? The answer to this question lies in the pedagogical model for calculating the current level of a student's knowledge which is used by the agent to display the module on the portal.

2.5.2 Designing Intelligent Agent Model

One of the most common solutions for the student diagnosis in ITS is testing. Generally speaking, test-based diagnosis systems use heuristic solutions to infer students' knowledge. In contrast, Computerized Adaptive Testing (CAT) is a well-founded technique, which uses a psychometric theory called Item Response Theory (IRT) [31]. The IRT supplies several methods to estimate students' knowledge. All of them calculate a probability distribution curve $P(\theta|u)$, where $u = u_1...u_n$ is the vector of items administered to students. When applied to adaptive testing, the knowledge estimation is accomplished every time the student answers each item posed, obtaining a temporal estimation. The distribution obtained after posing the last item of the test becomes the final student knowledge estimation. One of the most popular estimation methods is the Bayesian method [32]. It applies the Bayes theorem to calculate students' knowledge distribution after the posing an item i.

The students' knowledge level in an adaptive hypermedia application [33] is measured by using the answers to the questions previously presented to the student. The decision whether a student has solved the task presented in our proposed intelligent agent model is made on the basis of a formula which, aside from the correct answers, includes two additional parameters (time and the number of used Help options [34]).

In the ITS approach [35], user model content variables are used for keeping records on user interaction with the ITS and for adjusting the content presentation to the user profile. These learning style variables are a part of the Bayesian Network— BN for drawing conclusions about the student. There is a list of variables for each topic: spent time, topic depth level, wrong answers and correct answers. The variables relevant to deciding on students' knowledge in the GBMs system also include spent time and correct answers. However, the time needed for solving the tasks within our approach is divided into: (1) time used to read contents of the Help window and (2) time needed for giving the answers when the Help windows were not used.

The idea of the Neuro-Fuzzy Reasoner (NFR) system was the initial inspiration to create the model for students' knowledge diagnosis in a game-based learning system. The NRF system is relatively simple, supports creation of high-level pedagogical strategies, and can be easily adapted to individual teacher's preferences. The NFR model for student classification is based on test results and the time needed to complete the test. Modification of the NRF system parameters is made by adding a new parameter—time needed for reading the contents of the Help window. The learning model we have used in this education game is based on the operation principle of the NFR presented by Sevarac in his work [36]. The intelligent agent model has been extended with one more input variable—the Help window, because this component has been used by students in the educational game to a great extent. The initial model which we started with and which we used in the module N was based on the NRF.

Since the initial model had not produced the expected results, we applied the new model for knowledge level estimation which used the coefficients as variable values. The rule for determining the percentage of knowledge applies basic arithmetical operations to input parameters of the model. The significance of the input values for final knowledge estimation is determined on the basis of empirical coefficient values, given by the teacher. The knowledge level that student possesses after playing the education game is given by the following formula [37]:

$$P_i(X = \text{Mastered}) = \left(a \times \frac{A_i}{N} - h \times \frac{H_i}{N} - t \times \frac{t_a \cdot (A_i - H_i) - t_h \cdot H_i}{T_{\max}} \right) \times 100$$

$$(2.1)$$

where

- a, h and t are coefficients that should be estimated,
- A_iNumber of correct answers of i-th student,
- H_iNumber of opened help windows of i-th student,
- t_aAverage time a student needs to give an answer without using help window,
- t_hAverage time a student needs to give an answer when using help window,
- NTotal number of answers,
- $T_{\max}$Maximum duration of the game.

The values of coefficients: a, h and t can have a range from 0 to 1. The first part of the formula has the most significant role in calculating a student's final knowledge level, since it uses the number of correct answers—A_i. The second important part for knowledge estimation is the second part of the formula, which represents the number of used Help windows during the game—H_i. This part of the formula has the negative sign, since it decreases the probability of the final knowledge level. The part of the formula which depends upon time has the least importance for knowledge calculation. When a student uses the Help window, the time needed for giving an answer increases, which results in reduction of the student's total knowledge.

The basic terms of knowledge that a student has to present on a test are written in the help windows in the game. When a student starts learning by using the game or faces a difficulty during solving a task generated by the application, the Help window serves to accelerate finding the right solution related to a particular task. This means that an appropriate formulation of definitions and theorems within the Help window is the key moment in designing the entire application. There is a need to optimize the maximal duration of the game T_{max}. We have selected the approach to limit T_{max} by setting $T_{max} \geq N_{th}$. This admission is based on the assumption that when a student does not know the answer to a single question, he will use the Help window for each, namely Nth. The inequality in relation is set, because there is some additional time provided for a student to decide if he will use the Help window, (if he is not sure enough that he has a correct answer). If a student thinks that he has the correct answer to the query, he will provide an answer in shorter time ta and estimated knowledge level will be higher according to the Eq. (2.1).

Based on teachers' estimation, the values referring to the significance of coefficients in the given formula are estimated as:

- a—0.50,
- h—0.35,
- t—0.15.

where the sum coefficients must be 1. Through repeated comparison of the results obtained through classical paper test and results obtained through a computer game by using the estimation of the knowledge level with the new model (Eq. (2.1) and empirical coefficients) we have come to very encouraging results. With the use of the new model we have managed to reduce the error in estimating the students' knowledge level by 20 %. The new error value is 29.97 % ($\varepsilon = 0.3$) and it is reached by means of the above given formula and the empirical coefficient values.

The goal of the intelligent agent model is to estimate the knowledge of students as sufficient enough to let them pass the current level of module (sub-concepts) and go to the next one. Another goal is to compare and narrow the difference between the results obtained through playing a game and results obtained through classical examination methods. The results of classical examinations are graded into three groups: bad, good and excellent. Thus, we have graded the results obtained through using intelligent agent model in the same manner. The output of the model

estimates students' knowledge sorted into three grades: bad (P(X) ≤ 45), good
(45 < P(X) < 85) and excellent (P(X) ≥ 85). In our experiments we have
obtained results through classical paper-based tests that deviate a lot from the
results obtained through playing a computer game. For example, some students
passed the classical examination test with the highest scores (P(X) = 100) but they
answered less than 3 out of 8 level tasks in playing the game during maximum
time of game duration. Or quite the contrary—some students achieved very bad
results in classical examination tests (P(X) = 0) and they answered correctly to
more than 5 out of 8 level tasks in playing the game.

The selection of optimal coefficients is performed by using algorithm for
minimizing the root mean square (RMS) error [shown at Eq. (2.2)] between the
classical test results of students and the result of estimation model calculated by
the Eq. (2.1):

$$a_{opt}, h_{opt}, t_{opt} = \min_{\{a,h,t\}\in(0,1)} \sqrt{\frac{1}{N}\sum_{i=1}^{N}(R_{i\ classic} - R_{i\ est})^2} \qquad (2.2)$$

where R_i classic is result of ith student obtained by classical paper-based examination and $R_{i\ est}$ is result of i-th student estimated by intelligent agent model. The
C++ like code of algorithm implemented [38] is given in Fig. 2.7. At the output of
this algorithm the optimal values of a, h and t are calculated according to RMS
rule.

```
minError = 1000;
for(a=0; a<=1; a+=0.01){
  for(h=0; h<=1; h+=0.01){
    for(t=0; t<=1; t+=0.01) {
      ErrSum = 0;
      for (nStud=0; nStud<NSt; nStud++){
        Rest = (a*A[i]/N - h*H[i]/N - t*Ttot[i])*100;
        if (Rest > 85) Rest = 100;
        else if (Rest >= 45 && Rest < 85) Rest = 50;
        else Rest = 0;
        ErrSum +=(Rest - Rclassic[i])*(Rest -
Rclassic[i]);
      }
      Err = sqrt(ErrSum/NSt);
      if (Err < minError){
        aopt=a; hopt=h; topt=t;
    }
  }

  }
```

Fig. 2.7 C++ like code for calculation of optimal coefficients

The similar results may be obtained if other error criteria (like minimum average absolute error, maximal absolute error, etc.) are selected. The values of coefficients a, h and t calculated by the Eq. (2.2) are 0.55, 0.30 and 0.15. This leads to the error of 25.35 % ($\varepsilon = 0.25$). This is the main reason why the error of intelligent agent model estimation is even lesser.

2.6 Evaluation Results and Discussion

Modeling students' knowledge in educational games involves a high level of uncertainty. For this reason, the presented agent model could be applied through the pedagogical agent to game-based e-learning systems as a student knowledge diagnosis engine. Game-based learning represents one kind of software applications that uses games for the purpose of learning or education. The aim of such an application is to help the students understand the topics by visual representations of pertinent processes. In our evaluation we were interested in studying two factors:

- the search for the best coefficient values to make the model as precise as possible,
- how well the students learn the concepts from the GBMs environments.

Within our first study, we have analyzed the results achieved for each sub-concept through the classical paper-based test by the students who used the modules of our portal for the purpose of studying. The results of this test were collected in order to be compared with the results of students' knowledge obtained by means of intelligent agent model while playing the GBMs. The parameters were recorded during the playing modules of 71 students. The analysis has taken into consideration only the group of 50 students who used the modules in a time span lesser than maximum time allotted T_{max}. Also, this group did not include the results of the students who did not use the Help option while using the module itself. The reason for neglecting the Help option is the fact that the students who already had the knowledge about the given concepts used the module simply to practice what they had already mastered.

On the other hand, there are the students who wanted to finish the task in the module in the shortest possible time disregarding the potential inaccuracy of the given answers and thus ignoring the possibility of using the Help option. In order to calculate a student's knowledge by means of formula (2.1), the intelligent agent model in this study uses the following optimal values calculated by the Eq. (2.2): $a = 0.62$, $h = 0.25$ and $t = 0.13$. The mistake made by the agent model compared to the results of classical examinations on a paper test was very small $\varepsilon = 0.18$, as shown Fig. 2.8.

On the basis of the represented optimal values for the coefficients in this study the pedagogical agent has created the following relations in the knowledge

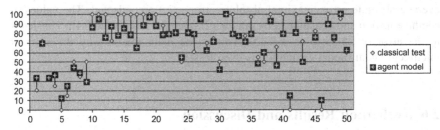

Estimated values by the agent model vs results from classical paper test

Fig. 2.8 The mistake made by the agent model compared to the results of classical test

assessment: the knowledge of 9 students has been equally estimated, the knowledge of 35 students has been underestimated while the knowledge of 15 students has been overestimated. Generally speaking, the results obtained through classical paper-based tests are better than values estimated by the agent model. In the intelligent agent model, each of the input parameters is weighted and the main task has been to calculate weighting coefficients in order to match the results estimated in this model to the results obtained through classical paper-based test. By comparing the results with the same set of the archive data, we also came to a conclusion that the values in formula (2.1) are partially sensitive to certain changes. The accuracy of the evaluation stays almost the same if, for example, we change the value of the rate t from 0.10 to 0.20 and the value of the rate h from 0.30 to 0.55. Slightly larger sensitivity of evaluation has been noticed while changing the value of the rate a. In order to maintain the model accuracy in a student's knowledge evaluation, the value of the rate a can vary from 0.50 up to 0.75.

Our second study dealt with the efficiency of the use of the GBMs as a supporting tool for preparing the exam questions in the field of the Z-buffer algorithm and the Unary logical operations. The total number of students attending was 183. However, the students that gave the "I don't know" answer were not taken into consideration in the analysis of the achieved results referring to this exam question. After the exam, we analyzed the results achieved by the students. We investigated the difference in the use of the GBMs between two groups of students, Group A and Group B. The group of students that used the GBMs for the purpose of the exam preparation was marked as Group A, while the other group that did not use the GBMs was marked as Group B. There were 82 students in each group. The exam results achieved by both groups are given in Fig. 2.9.

As shown in Fig. 2.9, Group A achieved better results than Group B. The difference in the number of correct answers is bigger than the difference in the number of incorrect answers. On the basis of these data, we can assume that Group A that used the GBMs was more successful than Group B that did not use the GBMs. To check this assumption we will use the method of statistical hypothesis testing in our research. In statistical research, the starting points are two mutually exclusive, opposite assumptions regarding the testing result—zero (H_0) and alternative (H_a) hypothesis:

Fig. 2.9 The exam results
achieved by both groups

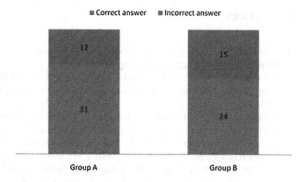

H_0 There is no statistically significant difference in distribution of correct and incorrect answers between Group A and Group B.

H_a There is statistically significant difference in distribution of correct and incorrect answers between Group A and Group B, and it is not accidental.

To determine whether the differences we observed within the comparative table are statistically significant (whether the distribution of values in rows and columns is independent), we used the X^2 test (Pearson Chi Square). The following values were obtained: $X^2 = 1.03$, df $= 1$, p $= 0.310$. From the Chi Square distribution table we read X^2 values for the selected significance level and the corresponding number of the freedom level. The corresponding probability X^2 (α) is 0.455. Since $X^2 > X^2$ (α) we automatically accepted the alternative hypothesis as the truth and concluded: the difference is statistically significant and it probably occurred under the influence of the experimental factor, which is in our case the GBMs. We have thus justified the use of the GBMs as a significant factor in the learning process. The presented results and conclusions drawn in this chapter attach considerable significance to the use of the GBMs in the fields that are abstract and difficult to understand, especially if we bear in mind that Net-generation students are not quite interested in the classic manner of learning. In that sense, the GBMs can represent good teaching material for other technical science courses as well.

Although we have done a lot of research on the intelligent agent model developed and used in the adaptive game-based learning, as well as virtual learning environment, our survey is not complete. A more thorough survey of the current trends in the applicability of knowledge management to e-learning system such as Moodle needs to be done. In the future we plan to create many GBMs for other units of Computer Science course and implement Moodle platform as classroom resources. In Moodle, any resource may be hidden by selecting option Hide from teacher. We have tried to develop Moodle plug-in based on the intelligent agent model which would automatically show the modules on grounds of estimated knowledge. The amount of knowledge P(X) necessary for students to reach the next module should be determined by the teacher. A large amount of data is to be gathered to find the optimal coefficients for the proposed model.

References

1. Shabalina, O., Vorobkalov, P., Kataev, A., Tarasenko, A.: Educational games for learning programming languages. System pp. 79–83 (2008)
2. Roslina, I., Wahab, S., Che Mohd Yusof, R., Khali, K., Jaafa, A.: Students perceptions of using educational games to learn introductory programming. Comput. Inf. Sci. 4(1), 205–216 (2011)
3. Callaghan, M.J., McCusker, K., Losada, J., Harkin, J., Wilson, S., Dugas, J., Demots, S., Desbois, F., Fouquet, A., Sauviat, F.: Game-based strategy to teaching electronic and electrical engineering in virtual worlds. International IEEE Consumer Electronics Society's Games Innovations Conference (ICE-GIC), pp. 1–8, 21–23 (2010)
4. Papastergiou, M.: Exploring the potential of computer and video games for health and physical education: A literature review. Comput. Educ. 53(3), 603–622 (2009)
5. Nguyen, L., Do, P.: Learner model in adaptive learning. Proc. World Acad. Sci. Eng. Technol. 35, 396–401 (2008)
6. Wooldridge, M., Jennings, N.R.: Intelligent agents: Theory and practice. Knowl. Eng. Rev. 10(2), 115–152 (1995)
7. Guttman, R., Moukas, A., Maes, P.: Agent-mediated electronic commerce: A survey. Knowl. Eng. Rev. 13, 147–159 (1998)
8. Shen, Z.Q., Gay, R., Miao, Y. : Agent-based E-Learning System—a Goal-based Approach, Business and Technology of the New Millennium, EdCTLeondes, Kluwer Academic Press International, vol. 3, Chap. 6, (2004)
9. Kelly, D., Tangney, B.: Adapting to intelligence profile in an adaptive educational system, Interacting with Computers. Elsevier 18, 385–409 (2006)
10. Villaverde, J.E., Godoy, D., Amandi, A.: Learning styles' recognition in e-learning environments with feed-forward neural networks. ISISTAN Research Institute, UNICEN University, Campus Universitario, Paraje Arroyo Seco, Tandil, Argentina (2006)
11. Burgos, D., Moreno-ger, P., Sierra, J. L., Fernández-Manjón, B., Specht, M., Koper, R.: Building adaptive game-based learning resources: The marriage of IMS learning design and <e-adventure>. Simul. Gaming 39, 414–431 (2008)
12. Jeetinder, S., Jayanthi, S.: Creating educational game by authoring simulations, Proceedings of the 17th International Conference on Computers in Education [CDROM]. Hong Kong: Asia-Pacific Society for Computers in Education (2009)
13. Weng, M.M., Fakinlede, I., Lin, F., Shih, T. K., Chang, M.: A conceptual design of multi-agent based personalized quiz game, advanced learning technologies (ICALT), 11th IEEE international conference on, pp. 19–21 (2011)
14. Conati, C., Gertner, A., Vanlehn, K.: Using Bayesian networks to manage uncertainty in student modeling. User Model. User-Adap. Inter. 12(4), 371–417 (2002)
15. Sabourin, J.L., Mott, B.W., Lester, J.C.: Modeling learner affect with theoretically grounded dynamic Bayesian networks. Proceedings of the 4th international conference on affective computing and intelligent interaction, pp. 286–295 (2008)
16. Li, Y., Musilek, P., Wyard-Scott, L.: Fuzzy logic in agent-based game design. Proceedings of the 2004 annual meeting of the North American fuzzy information processing society. Banff, Alberta, Canadá (2004)
17. Kuk, K., Prokin, D., Dimic, G., Stanojevic, B.: Interactive tasks as a supplement to educational material in the field of programmable logic devices. Electron. Electr. Eng. 2 (98), 63–66 (2010)
18. Kuk, K., Prokin, D., Dimić, G., Spalević, P.: Learning unary logical operations through the modern interactive educational application—Arhicomp. In: 10th anniversary international scientific conference, UNITECH'10, vol. 3, pp. 303–308. Gabrovo. Bulgaria (2010)
19. Tapscott, D.: Growing up digital: The Rise of the Net Generation. McGraw-Hill, NewYork (1998)

20. Felder, R.: Reaching the second tier: Learning and teaching styles in college science education. J. Coll. Sci. Teach. **23**(5), 286–290 (1993)
21. Norman, D.: Things that make us smart. Addison-Wesley, MA (1993)
22. Prensky, M.: Digital Game-Based Learning. McGraw-Hill, New York (2001)
23. Marlene, S.: The future: Gizmos. Technology and Children. J. Elementary School Technol. Edu. **13**(2) (2008)
24. Pitler, H., Hubbell, E., Kuhn, M., Malenoski, K.: Using technology with classroom instruction that work. ASC, Alexandria (2007)
25. Iqbal, S.T., Bailey, B.P.: Leveraging characteristics of task structure to predict the cost of interruption. CHI 2006 Proceedings: Using Knowledge to Predict and Manage. Montreal, Canada (2006)
26. Cox, C.T.: An investigation of the effects of interventions on problem solving strategies and abilities. (PhD, Clemson University) (2006)
27. Kuk, K., Rancic, D., Spalevic, P., Trajcevski, Z., Micalovic, M.: Use game based interactive multimedia modules to learning basic concepts on courses for computing science. Przeglad Elektrotechniczny **88**(5B), 150–153 (2012)
28. Parshall, C.G., Davey, T., Pashley, P.: Innovative item types for computerized testing. In: van der Linden, W.J., Glas, C.A.W. (eds.) Computerized adaptive testing: theory and practice. Kluwer Academic Publishers, The Netherlands (2000)
29. Kuk, K., Spalevic, P., Caric, M., Panic, S.: Game based learning module "Z-Buffer" on a course in computer graphics. In ProceedingsYUInfo 2011/ICIST 2011. Kopaonik, 6–9 March (2011)
30. Hwang, G.J.: A conceptual map model for developing intelligent tutoring systems. Comput. Educ. **40**(3), 217–235 (2003)
31. Lord, F.M.: Applications of item response theory to practical testing problems. Lawrence Erlbaum Associates, Hillsdale (1980)
32. Owen, R.J.: A Bayesian sequential procedure for quantal response in the context of adaptive mental testing. J. Am. Stat. Assoc. **70**(350), 351–371 (1975)
33. Font, M.J., Manrique, D., Ríos, J.: Evolutionary construction and adaptation of intelligent systems. Expert Syst. Appl. **37**(12), 7711–7720 (2010)
34. Kuk, K., Milentijević, I., Rančić, D., Spalević, P.: Pedagogical agent in multimedia interactive modules for learning—MIMLE. Expert Syst. Appl., Elsevier Sci. **39**(9), 8051–8058 (2012)
35. Fang, W., Blank, G.D.: Student modeling with atomic bayesian networks. Paper presented at the 8th international conference on intelligent tutoring systems. Jhongli, Taiwan, June 26, 30 (2006)
36. Sevarac, Z.: Neuro fuzzy reasoner for student modeling. IEEE computer society, Washington (2006)
37. Kuk, K., Spalević, P., Ilić, S., Carić, M., Trajčevski, Z.: A model for student knowledge diagnosis through game learning environment. Tech. Technol. Educ. Manage.—TTEM **7**(1), 103–110 (2012)
38. Kuk, K., Ilic, S., Spalevic, P., Panic, S.: Student knowledge diagnosis in game-based learning applications. In: 2012 IEEE 10th Jubilee International Symposium on Intelligent Systems and Informatics (SISY), pp. 455–459 (2012)

Chapter 3
E-Learning and the Process of Studying in Virtual Contexts

Dragoş Gheorghiu, Livia Ştefan and Alexandra Rusu

Abstract During the last several decades, the cultural pressures exerted by urban societies on the rural ones have led to a loss of cultural identity in the villages of South-East Europe. We believe that e-learning represents a solution for the preservation and continuation of the cultural identity in rural communities, as defined by traditional technologies and crafts. The present approach is based on a learning-by-playing teaching method and mobile-learning in real and virtual contexts. An original software application was designed for mobile devices making use of Augmented Reality technologies and delivering as main educational content 3D reconstructions of traditional environments and objects. The application integrates AI elements in the form of software agents. In this chapter the authors present the application prototype focusing on the role of software agents for the development of narrative e-learning tools and on the evaluation of the educational outcomes. Conclusions of this research and future work are also presented.

Keywords Cultural identity · Augmented reality · Software agents · Mobile-learning · Blended learning · Learning-by-playing

D. Gheorghiu · A. Rusu
National University of Arts, Bucharest, Romania
e-mail: gheorghiu_dragos@yahoo.com

A. Rusu
e-mail: rusu.alexandra.andreea@gmail.com

L. Ştefan (✉)
Institute for Computers ITC, Bucharest, Romania
e-mail: livia.stefan@itc.ro

M. Ivanović and L. C. Jain (eds.), *E-Learning Paradigms and Applications*,
Studies in Computational Intelligence 528, DOI: 10.1007/978-3-642-41965-2_3,
© Springer-Verlag Berlin Heidelberg 2014

3.1 Introduction

3.1.1 The Vanishing Identity of the Traditional Societies

A crucial problem of contemporary traditional societies is the protection and continuity of their cultural identity. The pressure exerted during the last decades by urban societies on rural societies has led many villages of South Eastern Europe to lose a significant part of their identity.

Starting from the premise that traditional crafts represent one of the crucial features of the identity of rural societies, it follows that their disappearance, due to the phenomenon of modernization, is particularly prejudicial to traditional societies.

3.1.2 A Brief History of Rural Romania

For centuries Romania was a country with a predominantly agricultural economy which, after World War II, was affected by a rapid process of industrialization and collectivization [13], changes which severely undermined the traditional rural society [52].

The period of the communist regime introduced more dramatic social changes, with even the creation of a new folklore, a new ideology, new forms and materials replacing the traditional ones, and the oral culture based on visual narratives, specific to the village mentality being replaced with a culture of the text, based on foreign models. This process continued during the last decades at an accelerated pace, with practically all the ancient traditions being forgotten within the span of two generations.

Vădastra, a village in Olt County, southern Romania, is a representative example that illustrates the crisis of the Romanian rural communities. It is for this reason that it was chosen as the place where the attempt to revitalize part of the traditional customs and technologies would be made, aiming to recreate with the help of the archaeological experiments, some of the ancient eco-technologies.

Initiated a decade ago, a series of research projects,[1] coordinated by Gheorghiu [18] from the National University of Arts Bucharest, succeeded in developing in Vădastra a small ceramics' center inspired from the ancient technologies. Aside from the initial scientific objectives, these projects also had strong educational goals, to be achieved by involving the young villagers in the process of recovery of their cultural past [47].

---

[1] 2006–2007, Grant CNCSIS 945, Space, water, fire. The Hydrostrategies and protechnologies of the traditional habitat, Romania; Gheorghiu [18].

The current research project conducted by Professor Gheorghiu at Vădastra—
"The Maps of Time—Real communities-Virtual Worlds-Experimented Pasts"
(Grant PN II IDEI) continues this trend, while attempting to transfer the educational content and programs resulted from experimental archaeology to the IT area,
with a strong focus on e-learning.

3.1.3   The Maps of Time Project

Our research is part of the project "The Maps of Time", Grant PN-II-ID-PCE-
2011-3-0245, in which both traditional and modern e-learning technologies were
developed. The latter insist not only on the identification, storing and transmission
of the data about traditional technologies, but also on the use for educational
purposes, of visual narratives about the ancient technologies, thus trying to revitalize a local custom dating back to the 18th century, that of painting visual
narratives with moral and educative subjects on the facades of various secular and
religious buildings (Figs. 3.1, 3.2).

Consequently, our interest was focused on a paradigm based on multimedia
visual narratives (also called hyper-story), implemented with the help of software
agents, and designed to develop the cognitive capacities of young people, while
simultaneously bringing an old tradition back to life.

After the successful experiment of revitalizing ceramic technologies in Vădastra,[2] we continued the process of recovering traditional crafts by focusing on local
textile production, a field of craftsmanship famous in the region in the distant past
[39]. Many of the old village women still practice the traditional craft, but this

Fig. 3.1 A narrative on the
ceramics' decoration

[2] 2000–2002, Gant World Bank and CNCSIS, The revitalization of ceramic tradition, Romania;
2003–2005, Grant CNCSIS 1612, The revitalization of traditional technologies, Romania.

Fig. 3.2 A visual narrative
written on the cultural
community center's façade

knowledge was not transmitted to the young generation, who ignore it completely. Therefore this subject was prioritized for our pedagogical activity in Vădastra.

The current chapter is structured as follows: (a) A section on related work; where related learning theories and implementation solutions are discussed highlighting the differences between the latter and our solution; (b) A section describing the agent-based learning paradigm, including a functional description of the overall learning solution and its components; (c) A section on implementation details focused on the software agents' functional role; (d) A section discussing experimentation of different scenarios of application use, presenting the learners' perspective of the e-learning solution; (e) A section on educational novelty and outcomes analysis; (f) Conclusions and future work, as the final section of this chapter.

3.2 Related Work

3.2.1 Theoretical Basis of the Educational Applications' Design

To deliver an educational message, teachers resort to modern pedagogical theories based on cognitive sciences and psychological behavior. The educational content has to be developed considering these theories, i.e. during a process named instructional design.

According to the instructional design theory, instruction should be organized in increasing order of complexity for optimal learning [40].

Robert Gagné [41] created an instructional theory that is currently used in the instructional design in different learning settings. One of his theories defines "curriculum as a sequence of content units arranged in such a way that the learning of each unit may be accomplished as a single act, provided the capabilities described by specified prior units (in the sequence) have already been mastered by the learner" [41]. In our case, this means the correct learning of fundamental stages of some of the traditional technologies which formed the identity of the Vădastra village (e.g. the textiles), represents a first stage of the learning process.

In [18] the author stresses that "learning software can be distinguished from pure information systems by the fact that its design is based on some conceptual models of teaching and learning." For the design of our educational software we considered theories of learning that applied to our paradigm.

3.2.2 Hypermedia Learning Environments

In [49] the author defines multimedia as a "new medium and a new communication technology" and he mentions that the "justification for learning with multimedia is that aspect of learning and instruction which may be designated as enrichment of learning." Schulmeister also considers multimedia "a second Gutenbergian revolution" [49]. In some implementations, the multimedia systems are complementary to the traditional learning style, while in others they are used as standalone systems aiming to completely substitute the professor's presence.

A hypermedia environment contains all kinds of media combinations: text, graphics, sound, and video. "Hypermedia is the integration of a computer and multimedia to produce interactive, nonlinear hyper environments" [49]. A hypermedia system resembles the human way of thinking, based on connections [18]. Some authors consider that it contains "the non-linear chaining of information which in a strict sense must exist in at least one continuous and one discrete medium" [53]. Schulmeister cites [36] as the one that defined the term of hyperlearning systems as "the combination of multimedia or hypermedia and learning; not a single device or process, but a universe of new technologies that both process and enhance intelligence." Schulmeister also states that a multimedia system must be "reactive, proactive and reciprocal interactive". It is important to remember that multimedia by itself cannot produce learning outcomes, as Clark emphasizes in [9], the latter being the result of the underlying instructional design.

Reciprocal interaction is implemented in modern computerized systems, like virtual reality, in which "learner and system may reciprocally adapt to each other" [49]. We conclude that a multimedia system is a highly interactive system. In such complex informational systems different modalities and techniques are used to inter-connect the constitutive media, e.g. system-user interactivity, narrative structures, artificial intelligence or software agents, which ensures the "navigability, adaptivity, reactivity" of these systems [49].

Regarding the modern cognitive theories and learning and teaching paradigms, the multimedia learning systems support an autonomous self-paced learning style and a constructivist learning paradigm, which is a form of active learning, or "learning by doing". According to this model, the learners develop their knowledge in a participative way, gaining a practical experience in real situations or in environments resembling the real ones [14]. The active learning is typically implemented with environments featuring a kind of immersion. The hypermedia systems also support the non-linear learning, and try to recreate the spontaneity of the learning process. The learner is presented a multitude of learning opportunities and subjects from which he can select the desired ones based on individual preferences, level and educational necessity [14].

In our research project we have attempted to implement a modern paradigm of learning traditional technologies within their genuine historical contexts, based on the local tradition of visual narratives, an educational strategy tailored for children between the ages of 8 and 12 years old. We surmised that the transformation of simple visual narratives, describing the technical processes, into digital stories would produce an efficient tool of mobile and contextual learning for children.

Since the stories are "narratives of true or fictional events that intend to capture and involve learners actively" and have "a topography, and spatial and temporal dimensions" [43], they were ideal for our contextual approach. A special type of narrative is the hyper-story, based on multimedia content and able to develop "cognitive structures that determine tempo-spatial relationship and laterality in early age children" [43]. Through this description Sanchez and Lumbrera emphasize the importance of spatial and temporal features in the process of learning, or, in other words, the importance of the receptor's situation, an aspect which we took into consideration in our approach. By exploring the environment children learn in a process similar to a play; this is why our paradigm could also be described by the term "learning-by-playing".

3.2.3 Mobile Augmented Reality

As a personal learning style, mobile learning is defined as "the acquisition or modification of any knowledge and skill through using mobile technology, any-where, anytime and results in the modification of behavior" [15]. Mobile learning also supports the "situated or context-based learning" model [1], which stresses that learning has to take place in "true learning contexts", and contain "authentic activities and assessment" [12].

In order to implement a mobile-learning system with hypermedia simple or complex narrations, we considered the technology of the Augmented Reality (AR) on mobile devices.

Augmented Reality is a computer technology which allows the dynamic overlapping of multiple computer generated content layers, on a live-view camera stream, while tracking the user's movements and view changes. AR is seen as supporting a new immersive visualization paradigm and a natural human-computer interaction. It is for this reason that AR technology is adopted in educational projects, to mediate an enhanced visual and cognitive perception, and to support several new learning styles: mobile, situated in context, ubiquitous and nomadic, social. The content may comprise texts, 2D images and graphics, 3D graphics and animations, audio and video files superimposed on the surrounding reality, captured by means of a video camera, or optical systems (HMD or glasses). These augmentations have to be integrated naturally and in real time on the video stream represented by the real scene, making the user perceive a new cognitively augmented scene. This is achieved by means of complex and advanced techniques, such as real-time tracking and 3D registering. The technical definitions of AR processes can be found in the reference paper of Azuma [2].

Some authors [40] consider that video information does not represent an augmentation of reality. When these augmentations are visualized outside the AR process, they are referred to as "actions" attached to AR, but when integrated in the real video stream using video-in-video techniques they are part of the AR process. The use of the video augmentations has great pedagogical value, and opens the way to integration between AR and multimedia systems, leading to a new form of hypermedia-learning system, the generic Mixed Reality (MR) system [11].

The AR/MR systems allow the augmentation of the user's perception of the real world, but also the use of existent visual and spatial abilities and an enhancement of the interaction capacities of the users [19]. Specifically, compared to MR systems, the AR applications allow the implementation of context sensitive applications. When the context is geographic, the AR applications act as true information browsers. These kinds of AR applications present a non-linear navigation metaphor for content visualization and make use of remote data services. The augmented data delivery is determined by certain "triggers". The similarity with web applications is obvious, except that the data request is triggered not by the user, but by a geographic context. From this architecture we conclude that AR applications are strongly dependent on online connections, a feature which can affect the usability of AR applications for learning purposes.

Consequently, our approach is to develop an AR learning tool as a native application for Android i.e. to use integrated hardware and software capabilities to store both the application and the content. The native application approach has the disadvantage of being restricted to specific mobile platforms.

The present AR technology allows a seamless integration with social networks by means of service mash-ups. In our approach we used combined software agents with social network mash-ups.

Having thus described the AR processes' capacity to augment a user's understanding of reality in its context, we must now question whether, for an educational application, a simple visualization in an AR view is sufficient even if it is ideally implemented (as defined by Azuma [2]: 3D content overlaid in real time on the live video image). We consider that this is sufficient only for achieving a first cognitive level but not for a learning purpose.

That is why we proceeded to investigate technologies to be associated with AR in order to develop an educational application. Furthermore, targeting a group of primary and secondary school children, this kind of application needs to include elements to engage this category of learners and to offer a certain level of application control. Analyzing modern software engineering technologies we considered software agents, which are discussed later in this chapter.

3.2.4 Agent-Based Approaches

The learning process is a dynamic one and traditionally mediated by the teacher. Modern models of e-learning, are focused on a learner-centered model, and pedagogically mediated by technology. Conrad and Donaldson [10] define several methods for creating engaging courses in the e-learning environments.

Software agents as well as AI (Artificial Intelligence) are used in educational software to create intelligent applications, adaptive to the context of their use, to support this learning paradigm and also to capture the learner's interest.

Narrative based educational application use software agents under the form of animated humanoid agents, namely in the form of characters, which are designed with the purpose of rendering the lesson more engaging, to compensate for the absence of the teacher, or to provide help to students in different situations occurring during the educational process. The agents guide the learner through the learning process according to the pedagogic strategy.

In multimedia systems, agents serve as "personal metaphor guide (guides, agents, tutors) or historical personages" [49]; they make the connection between the learner and the educational system. An example is the one implemented by Sanchez and Lumbreras [44] and Sánchez et al. [43].

Most implementations of software agents represent complex applications with elements of artificial intelligence, e.g. digital storytelling, collaborative, gaming. Software agents were extensively used in Virtual Reality based educational applications having visual interfaces.

In [4] the authors perform a brief review of software agent implementations in AR environments, as these are more recently integrated within AR environments. An early AR application with humanoid agents is the ALIVE system [37]. In [8] a live video avatar of a real person is placed into a Mixed Reality setting, and interacts with a digital storytelling system with body gestures and language commands. In [3] are experimented interaction techniques with virtual humans in Mixed Reality environments, which played the role of a collaborative game partner and an assistant for prototyping machines. For [7] the agent is a virtual playmate assisting children in a natural storytelling play with real objects.

In [19] the authors review implementations of Mixed Reality applications which use the technologies of software agents for the control of the actions/tasks and also as a user interface metaphor.

Archeoguide [17] is an example of an outdoors AR application that uses X3D data format and runtime to describe the content and a dynamic runtime behavior, which contains elements of AI logic written in JavaScript. An X3D scene consists of three different layers: background video, 3D reconstructions and the user interface [17].

AMIRE it is an authoring language developed within an EU IST Program, completed in 2004, dedicated to efficient creation and modification of Mixed Reality (MR) applications, authoring metaphors, and generic design recommendations and procedures [20].

As we can see, state-of-the-art approaches exist for different educational purposes and learning settings. Of these, only [17] and [34] use software agents in mobile AR settings, similar to our approach. In [17] a proprietary AI solution was developed for an intelligent and adaptive touristic guide. In [19] open technologies were used for an educational application using visual agents which assist the learners. In our approach we used non-visual software agents as AI components in order to support a narrative learning application with both pre-determined and non-linear learning paths. For the implementation we chose both open source and commercial SDK.

3.3 Description of the Agent-Based Learning Paradigm

3.3.1 Motivations for Present Work

We believe that e-learning can be a solution for the support of the preservation and continuity of many cultural traits of rural societies, such as textile technologies, ceramics technologies, as well as various other crafts.

In our case of vocational learning the successful integration of new e-learning technologies is due to the application of a blended learning strategy, i.e. a combination of traditional classroom teaching (or workshops), and e-learning specific technologies, to cite only the videoconference interactive system, or the use of virtual learning environments or mobile learning. This strategy is described by Schlosser and Burmeister [48] as using the "best of both worlds".

To kickstart this pedagogical project of the revitalization of the traditional textile technology, the local school was equipped with a Sony Bravia monitor with a CMU-BR100 video camera, and with Skype Internet connection. The first video conferences were conducted in the summer of 2012, with the courses delivered by the staff of the Textile Department of the National University of Arts Bucharest and attended by both village teachers and students. In the beginning, face to face hands-on lessons (consisting of presentation of techniques and individual work with groups of village students) were performed using replicas of Roman looms.

In the subsequent stage the project set the basis for the development of a community of learning which advanced the blended-learning approach by creating a platform of experimentation of the technologies, as well as a virtual and inter-active e-learning system.

To support a new paradigm for informal learning, as a precursor to further vocational training, an original software application was designed for Android mobile devices (smartphones and Tablet PCs), integrating at least a GPS receiver and a rear video camera.

The application assists children as they progress through each lesson in a mobile real/virtual environment based on the Augmented Reality technology and the concept of geo-referenced Points of Interest (Figs. 3.3, 3.4). Thus, the application is dependent on a well-defined geographic context.

Fig. 3.3 The application and the list of POIs

Fig. 3.4 The mobile AR application

The educational content is structured under the form of hypermedia narrations (as individual lessons), consisting in walkthroughs within a multimedia interconnected content. Each lesson uses 3D reconstructions of the traditional objects and of the original environments in which these objects were used. The children select a specific digital narration according to their learning objectives. The whole application integrates several specialized functional modules including a framework for software agents. The application was developed following the Android application model.

3.3.2 General Presentation of the E-Learning Solution

In our research project we have attempted to implement a modern paradigm of learning traditional technologies within their genuine historical contexts, based on the local tradition of visual narratives, an educational strategy tailored for children between the ages of 8 and 12 years old. We surmised that the transformation of simple visual narratives, describing the technical processes, into digital stories would produce an efficient tool of mobile and contextual learning for children.

Since the stories are "narratives of true or fictional events that intend to capture and involve learners actively" and have "a topography, and spatial and temporal dimensions" [43], they were ideal for our contextual approach. A special type of narrative is the hyper-story, based on multimedia content and able to develop "cognitive structures that determine tempo-spatial relationship and laterality in early age children" [43]. Through this description Sanchez and Lumbrera [44] emphasize the importance of spatial and temporal features in the process of learning, or, in other words, the important of receptor's situation, an aspect which we took into consideration in our approach. By exploring the geographic and historic environment, children learn in a playful manner, searching for Points of Interest (POIs) representing different learning stages; this is why our paradigm could also be described by the term "learning-by-playing".

Following the choice of implementing our solution on mobile equipments (smartphones and tablets) we promoted the mobile learning paradigm.

The programmatic power of the software agents' technologies was leveraged in order to implement a multi-agent system. One of these agents is the "learning agent", which has the main task of mediating the learning process. The agent-based learning solution will be described in the following sections.

The learning paradigm experimented in our project consists of a hypermedia learning system in the context of an Augmented Reality—based environment. The software application provides a logical suite of stages of learning of the traditional technologies (textiles, glassware), while leaving the children the freedom to choose a learning path, i.e. a specific lesson. By using our application, the children can see all the stages of a technology, both as a pure technique and a technique in a cultural and historical context.

This educational scenario is addressed to children and contains elements to attract and persuade them to travel through a complete lesson. If this does not happen, i.e. a lesson is not completed, this status is persistently stored, such as at the next use the application is able to make the recommendation to resume the lesson from where it left off.

For our paradigm we therefore needed to implement scenarios based on multimedia sequences and an application to provide intelligent, adaptive behavior. This required an implementation of methods for monitoring the learning process to provide feedback on the application usage, and a behavior adapted to the context of use. Instead of traditional AI programming, we chose to use software agents, due to their ability to incorporate both logic and communication mechanisms, and therefore adequate for designing very specialized functions.

3.3.3 Basic Concepts of Software Agents

In computer science software agents represent intelligent software modules, which can act in the name of the user, or other software module by means of an agency. This represents "the authority to decide which, if any, action is appropriate" [38].

In the domain of Human-Computer Interaction (HCI), the agents are referred to as interface agents or user agents [35].

Software agents also represent a modern programming paradigm [50], Agent-Oriented Programming (AOP), different from the procedural or objectual ones, but close to the concepts of methods and functions.

There are several definitions of the software agents: "persistent software entity dedicated to a specific purpose" [51]; "self-contained program capable of controlling its own decision making and acting, based on its perception of its environment" [55]; "effective for organizing and programming software applications in general, starting from those programs that involve aspects related to reactivity, asynchronous interactions, concurrency, up to those involving different degrees of autonomy and intelligence" [45].

According to [55] a software agent is defined as a hardware-software entity having the following essential attributes:

- *Autonomy*: agents can operate without the direct intervention of humans or others, and have control over actions and internal states;
- *Social ability*: agents can interact with other agents, as well as with their users;
- *Reactivity*: Agents can perceive their environment and respond on time to the changes;
- *Pro-activity*: agents can take the initiative and display a behavior based on an objective.

This autonomy allows agents to accomplish tasks which are defined in terms of "behaviors". Because of their autonomy, software agents can be taught by high-level descriptions so that the user is freed from complex low level operations. With this programming model difficult programming problems are solved such as concurrency, security or context-sensitivity [45], and also predictive agents can be developed.

There are several open source agent-based technologies on mobile devices, such as JADE [21], JaCa [22], Jadex Android [23], Agent Factory Micro Edition [24], 3APL [25].

In [45] is presented the JaCa-Android as an integration of two existing agent programming technologies: Jason and CartAgO. Jason is an agent programming language based on AgentSpeak language, the most employed agent-based language, and also a platform [26] developed in Java. CartAgO is a framework for programming and running the agent-based applications [27]. We took into consideration this platform for our agent-based implementation, due to the fact that it is a high-level framework and it provides an Android version.

Jason's architecture is based on the BDI (Belief-Desire-Intention) computational model, which defines the behaviour of individual agents and offers a model for agent reasoning. The agents can be reactive (i.e. react to events) or pro-active (try to achieve a desired status). On the environmental side, CArtAgO uses the notion of artifact as an abstraction to define the structure and behaviour of environments and the notion of workspace as a logical container of agents and artifacts. Artefacts represent the environment resources and tools that agents may dynamically instantiate, share and use [45]. The advantage of JaCa-Android is that it provides a high-level developing framework based on abstractions that facilitate the development of agent-based mobile applications on a mobile platform [45].

The advantage of the programmatic power of the software agents is counterbalanced by the disadvantage of their complexity and steep learning curve. A high-level framework allows a software developer to concentrate on the design of the application functionalities, contributes to a lower development time and is less error prone. Nevertheless, there still remains the issue of understanding which types of software agents and development framework are suitable for a given application and the need to allocate time for an evaluation of the existing frameworks as there are several frameworks offered by an open-source community, at different development stages.

3.3.4 Components of the Agent-Based Learning Paradigm

The paradigm we propose has the following structure:

1. An *informal learning-by-playing* and mobile paradigm focused on skills and technology;
2. A situation in context and information visualization using Augmented Reality technology;
3. A digital narrative using hypermedia (i.e. linked multimedia content);
4. Sensitiveness to the context and usage with the help of software agents, i.e. dynamic adaptation of the didactic content and of the application interface.

In the present chapter we will discuss only the learning of a single craft technology, that of traditional textiles. It is well known that textiles, as well as ceramics, was a conservative technology [5], remaining almost unchanged during millennia. Consequently, the methods of weaving documented in Vădastra village date back at least from Roman times. This rationale combined with the fact that Vădastra, like many other villages in the region, was built overlapping a Roman *villa rustica*, prompted us to design the lessons on textiles with historical references, thus allowing the children to experience a lesson of history, beside the technological lesson.

As a result, the process of learning proposed by us is based on the concept of a Hypermedia Digital Storytelling Scenario (HDSS). A HDSS could be defined by the following formula:

the dynamic content (the 3D architectural reconstructions + the 3D technological objects reconstructed) + the story of technology + the performance of the experimentalist (Fig. 3.5).

The educational application designed for this purpose, named MapsofTime-LearningTool, offers the following 3 learning levels, presented in order of their gradually increasing cognitive complexity.

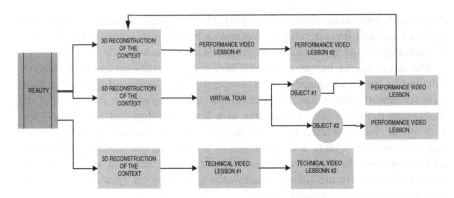

Fig. 3.5 Hypermedia digital storytelling graph (UML diagram)

1. The first stage begins with the *Context's Digital Storytelling,* or, in other words, the visual story of the place, which is a narrative about Romans and their technologies, positioned via GPS within the local geography, and presented under the form of video films with the re-enactments produced by the team of experimentalists of Vădastra. For this purpose a couple of horizontal and vertical Roman looms [6, 54] were reconstructed. To increase the veracity of the technological performance, performers were dressed in Roman costumes and acted in a reconstructed *villa rustica*'s workshop and outdoors (Figs. 3.6, 3.7). Thus children were able to visualize the textile technology in its ancient context, a historic information layer enhanced by the performance of the contemporary technologists.

2. The second stage is the *Objects' Digital Storytelling,* in which the 3D reconstructions of the objects for weaving,[3] furniture (Figs. 3.8, 3.9) and interior

Fig. 3.6 3D reconstruction of a Roman *villa rustica* (*Context's Digital Storytelling*)

---

[3] 3D reconstructions by Assis. Prof. Alexandra Rusu (National University of Arts, Bucharest).

Fig. 3.7 Artist showing weaving techniques (*Context's Digital Storytelling*)

design details were included. Children can play with the animated virtual reconstructions, understanding the meaning and function of the displayed objects.

3. The third stage, the *Technology's Digital Storytelling* was designed for those who acquired a high level of technical knowledge, since it presented in detail the processual stages (*chaînes-opératoires*) of the technology studied (Fig. 3.10). The children truly interested in learning the technology can access this level of the hyper-story made of video films describing the technical stages,[4] with images captured from both the observer's and the performer's perspectives. To contextualize these operations some images had links to the first stage of the hyper-story.

Fig. 3.8 3D reconstructions of the objects (*Objects' Digital Storytelling*)

---

[4] Video footage: Lect. Adrian Şerbănescu (University Spiru Haret, Bucharest).

Fig. 3.9 3D reconstruction
of a Roman vertical loom
(*Object's Digital
Storytelling*)

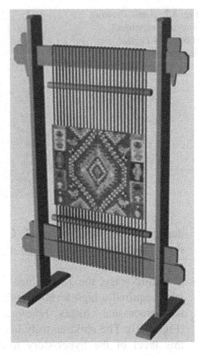

4. The practical part of the hyper-story, the *Craft Approach,* was materialized
   under the form of workshops organized for the village children,[5] where they
   worked with the reconstructed Roman looms (Fig. 3.11). This was a good
   opportunity to develop their skills and discover real talents. At this stage too,
   children can access the first stage of the hyper-story.

Fig. 3.10 Textile
technological gestures
(*Technology's Digital
Storytelling*)

---

[5] Weaving experiments: Assist. Prof. Alexandra Rusu and technician Elena Haut (National
University of Arts, Bucharest).

Fig. 3.11 Hands-on lessons
with children about weaving
techniques using a vertical
loom (*Craft Approach*)

In the hyper-story we designed, the child has limited decisional control; s/he can choose a story, for example the visual story of the place, but cannot intervene in the (pre-determined) order of the technological stages. On the other hand, the child has the freedom to combine the stages of learning in a real-augmented environment with the practical experimentation of the technology.

3.3.5 Description of the Software Agents

In order to support the concept of HDSS, we implemented the following categories of software agents (Fig. 3.12):

1. User-Profile Agent (UP-A)

The UP-A agent manages the user profile by name, associating it with the result of the current lesson completion status, in relation to the chosen technology (textile, glass, ceramic). This agent delivers the welcome message and also transmits information to other agents (the learning agent, the user interface agent).

2. AR Context Agent (AR-A)

The context can be considered an agent for information filtering, making the application context sensitive. To support this, a geographic area was predefined in order to contextualize the learning solution. The area represents at large an archaeological location from Vădastra village. In his area, for each of the taught technologies, 3 geographic points of interest (Geo-POIs) were defined by their coordinates, representing the three learning levels (according to their description in Sect. 3.3.4).

The geographic context is given by the GPS, compass and accelerometer sensors of the mobile devices. In the AR-A agent the AR functionality and logic were embedded: receiving and interpreting the information from the sensors; deciding first if the user is in the pre-defined area; only in this case and according

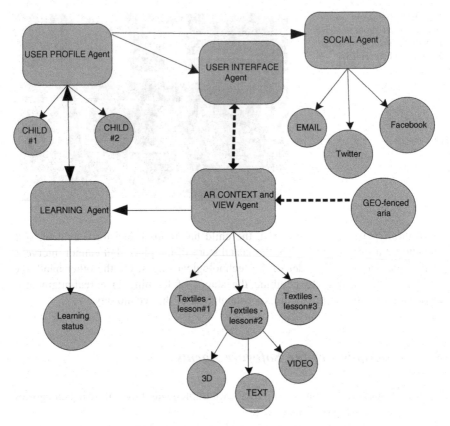

Fig. 3.12 The agents and a diagram of their inter-communication

to the chosen technology, the three POIs will be displayed. For each selected POI, the AR software overlays augmented information over the video stream, and ensures the 3D registration and tracking processes.

The AR-A agent also manages the hypermedia links inside of each learning level and communicates with LE-A, the learning agent.

3. User Interface Agent (UI-A)

The interface displays the POIs to which the narrative lessons are attached and allows a non-linear (at choice) navigation among different learning levels related to the technology in context (e.g. textile technology). The geographic points of interest (Geo-POIs) are displayed either on a Google map or in a list (Fig. 3.3), and have attached colored labels with information.

The UI-A agent adapts the interface according to the user's profile, making use of the messages received from UP-A. For example, the POIs representing learning levels completed by a particular user which returns to the application will be displayed with a different color background.

4. Learning Agent (LE-A)

This agent supervises the learning process: the agent receives message from the AR-A agent regarding the completion status of a particular learning level, and transmits specific messages to the UP-A agent. When the child returns to the application, he is free to choose a learning level, but he also receives a status information or recommendation due to collaboration among AR-A, LE-A and UP-A, UP-A and UI-A agents.

The application usage information monitored by LE-A can be used for further investigations, e.g. lesson completion rates.

5. Social Agent (SL-A)

This agent implements the connections with social networks. The learners can share their experience by sending an image and text by email, to a Facebook or Twitter account. The SL-A agent integrates the underlying service mash-up, by implementing an OAuth authorization service [28].

The control of one *learning level* comprising several digital stories, i.e. the link between the multimedia components in a learning scenario, occurs in a linear manner for users, in order to direct them through a thread of a story. The scenario follows the template:

- The real environment is assigned the 3D reconstructions;
- The 3D reconstructions are followed by presentation of 3D objects;
- The 3D objects are followed by video presentations.

In the AR view of the application, the navigation through an individual learning level is achieved by means of interactions with the virtual objects, i.e. attached events which serve to advance to the next element of the story.

3.4 Implementation Details

3.4.1 Software Integration Challenges

Our learning application is designed to combine several software technologies: AR, multimedia content, 3D content, software agents on Android mobile platform, social media.

Software agent technology was chosen as a solution to provide specialized software components in order to manage the learning application as a whole. The existing AR platforms can manage the AR processes and the social media integration, but cannot provide support for managing a complex custom application with AI behavior.

From the software implementation point of view we had the choices, according to [45, 46] to either port agent-based software technologies on mobile devices or use frameworks developed for mobile devices. To integrate a software agent framework we needed to select an AR development framework not an AR platform which acts as Content Management Platform.

3.4.2 The Android Platform

The AR application was designed for Android smartphones and tablet PCs. The integration of different functional modules was possible by developing a native (device-specific) applications on the Android platform. This offers an open source software stack for mobile devices which encompasses the operating system (Linux-based), native libraries, Java runtime (virtual machine), the Android application framework (SDK) and user applications. This open architecture allows for modularization and facilitates the development of complex applications.

The Android SDK is an object-oriented Java-based framework, consisting of a high-level set of classes, providing abstractions for developing mobile applications for Android-enabled mobile devices. The main components of the framework are:

- Activity Manager: controls the life cycle of all Android applications;
- Fragment Manager: manages user interface elements (e.g. a list of menu items);
- Services: a service does not have a GUI and runs in the background for an indefinite period of time;
- Content providers: manages specific applications' data that can be shared with other applications. The data can reside in an Android SQLite database or XML files;
- Resource Manager: manages application resources, such as literal strings, images, XML files;
- Location Manager: provides location information related to the device;
- Notification Manager: allow application to notify the user regarding different events.

The Android framework also uses low-level synchronization mechanisms in case of complex applications with asynchronous events. This behaviour is similar to thread-based programming used in other programming languages. The agent programming model also uses asynchronous communication, but managed through an agent-based framework, which abstracts the low-level platform communication implementation.

3.4.3 The Mobile Augmented Reality Application

For the development of a Mobile Augmented Reality application, three options are available:

1. First, and the most common, use of a client-server architecture, which involves the AR platform, the application developer or third-party server for content management and a client software residing on the mobile device and downloadable for different mobile platforms; in this case, the content delivery is dependent on the internet connection.
2. Second, is to develop a native application for a specific mobile platform, package it with the content and install on a mobile device;
3. The third, is a combination of the first two options: the native application and the content are installed on a mobile device, and those can be upgraded from time to time on an internet connection.

We chose to build an application following the third architecture model, i.e. a native applications less dependent on an internet connection bandwidth and availability but using it for periodic application and multimedia content update. The development of a native application enabled us to seamlessly integrate the software agent framework with the AR application and ensured an optimal performance, but have raised the complexity of the application design, implementation and test.

Three commercial AR platforms were analyzed: Layar [30], Wikitude [31] and Junaio [32], with the final selection being Junaio from the Metaio company. The Metaio AR SDK offers more options for developing geographical AR applications using a modern, powerful and flexible programming model (Fig. 3.13), allows development of interactive scenario-based AR applications and supports multiple tracking models (optical and non-optical).

Regarding the authoring of the Augmented Reality scenarios for each individual multimedia story, i.e. learning level, we used UML (Unified Modeling Language) diagrams (Fig. 3.5) and Metaio Creator [29] which is a visual authoring tool. The application was then deployed to a native Android application using the AREL export function.

Augmented Reality Experience Language (AREL) is the JavaScript binding of the Metaio SDK's API, which uses a static XML content definition. AREL is a scripting-based programming model which can be employed for highly interactive Augmented Reality experiences, in conjunction with open web technologies such as XML and HTML5.

The Metaio SDK also allows the multimedia components for an individual scenario (learning level) to be defined in the XML file. After the application is loaded on the device, these components are stored locally, so the Internet connection is necessary only for updates.

The Android application model uses views for each information screen. Our application uses 2 main ones—MetaioSDKViewController and MetaioSDK-ViewActivity, which contain all the system calls required to control the augmented reality experience (e.g. open the camera). These custom classes can be extended and overwritten for a custom application logic.

To delineate the geographic area of interest, we applied the geo-fencing technique: the area of interest being limited to the perimeter corresponding to the

Fig. 3.13 The architecture of Metaio's development tools [32]

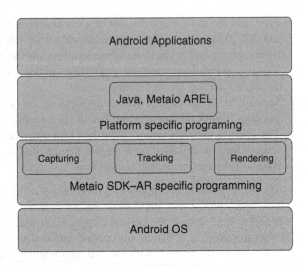

Roman villa archaeological site. Once the user exits this area, the application no longer provides points of interest.

The POIs are marked using the billboard concept, i.e. a pop-up with a background image, icon and text.

3.4.4 The Multimedia Educational Content

The studied context, i.e. the Roman *villa rustica* and its objects, were modeled in Autodesk 3ds Max[6] and used in applications and MD2 Wavefront OBJ formats. This latter format allows animations, e.g. rotations, etc., and the 3D objects contain geometry and textures packed in zip archives. The virtual tour through the courtyard was also implemented with the help of 3ds Max, and contains renderers in a MP4 compressed video format.

All the digital assets (images, videos, 3D models) were packaged in small files (under 5 Mb), according to the constraints imposed by the mobile devices, and also for efficient transfer over an internet connection.

Our multimedia, superimposed on an AR scene, is based on several digital assets: images, videos, 3D models. We authored the three multimedia hyper-stories using Metaio Creator which can implement an interactive multimedia scenario, e.g.: playing full screen (streamed) videos; showing/playing any content (2D, 3D, video, sound) in AR mode based on the supported tracking technologies; starting, stopping, looping animations, videos, sounds; moving, rotating, scaling any content (2D, 3D, video); automatic content adjustment to the tracking object. In order

---

[6] 3D reconstructions by: arch. Andra Jipa, Cristina Kudor and Inga Bunduche (National University of Arts, Bucharest).

for the content to be 2D/3D registered on the real life target scenes, several manual calibrations were necessary.

3.4.5 Social Media Integration

In Android, interactions among components are managed with a messaging mechanism based on the concepts of intent and intent filter. An application can request the execution of a particular operation offered by another application or component, by providing to the O.S. an intent with the information related to that operation. The O.S. will handle this request locating a proper component—e.g. a browser—able to manage that particular intent. The intents manageable by a component are defined by specifying a set of intent filters.

The connection of the learning application to web sites, email, Twitter and Facebook accounts is done using Android intents and is implemented using an OAuth authorization service, as user authentication is required to access these networks. This process is also called service mashup.

3.4.6 Software Agents Implementation

Our main concern in using software agents was to implement a mechanism to support a flexible behaviour of our application designed to be used for learning purposes. Instead of incorporating AI elements or other logic control we considered the software agent paradigm on mobile platforms more challenging. We designed software agents having specialized tasks to perform, capable to ensure the overall functionality of the application and provide assistance on the learning process. We mainly leveraged the communication mechanism among software agents, which we considered very adequate for our task. We did not implement self-learning mechanisms or other advanced AI functions, but this can be done further using the existing software agent programming model.

Regarding the BDI (Belief-Desire-Intention) architecture, we focused on programming plans for agent's "intentions". Intentions represent a deliberative state of the agent, which is translated in the execution of a plan, i.e. sequences of actions performed by an agent to achieve one or more "intentions".

For software agent programming we have selected the JaCa—Android framework [55], with which we programmed several software agents characterized by a clear underlying logic and communication channel among them.

Using JaCa we programmed Jason agents with the logic to execute tasks required for our mobile learning application with the purpose to:

1. Manage the overall application;
2. Make the application adaptive to the usage scenarios;

3. Implement asynchronous communication mechanism among different application components, i.e. agents having different roles and tasks, by programming a reactive agent behaviour.

Navigation in the hypermedia space is performed on two levels:

1. A simple interaction with the digital augmentations in the AR view, which are treated by the Augmented Reality agent (AR-A), according to the scenario developed using Metaio Creator software utility and further customization. This provides a predefined individual navigation through multimedia story elements;
2. Navigation supervised by the learning agent (LE-A), which detects scenario completion status and informs other agents about this condition, which can further assist and inform the learners about their learning process status.

The first level is based on UI events and on an XML list containing digital assets, and associated to the Geo-POIs. The other level uses specific software agent communication and logic.

The learning agent (LE-A) receives a status message regarding the lesson completion, and together with the User-Profile agent can construct a learning profile of the current user and generate specific messages. For example, detecting that a user repeatedly does not complete a learning path, will recommend him either to go through the entire lesson or to chose another learning level which suits him.

The LE-A agent can also store data regarding the application usage in an Android SQLite internal database. This data can be extracted from the device at a later time for analysis of the learning process with our application.

3.5 Application Experimentation

The children were involved in the testing phase of the application. They used Android smartphones and PC tablets (Fig. 3.14) provided by the research project.

The start page of the AR application allows user registration and selection of one of the technologies (at this application stage, textiles). The registration is optional; in case it is overridden, the user profile agent (UP-A) manages the application using an anonymous user account.

The selection of the technology (i.e. textiles) acts as a filter for the displayed Points of Interest (Fig. 3.3) corresponding to associated learning levels (L#1, #L2, #L3). A POI selection further triggers the start of the hypermedia story, consisting of linked multimedia augmentations over the AR view. The scenario completion is supervised by the learning agent (LE-A).

The children were presented the usage scenario: upon opening the application and inserting their names they received a welcome message and a history of their use of the application. When approaching the areas of interest, in this case the Roman dwelling, the application displayed the geographic POIs and the children

Fig. 3.14 Children using the tablet PC version of the application

could use radio buttons to select the Roman technologies (currently TEXTILES, CERAMICS and METAL, in the future versions). When the children left the GPS selected area, this information disappeared from the application's display. Therefore, through a series of trials the children would situate themselves inside the perimeter of the Roman dwelling, this situational game being the first educational stage of the application. The children then chose a technology to learn about, for instance "TEXTILES", using the touch-screen display. This triggered the display of the specific lessons under the form of billboards floating on the screen. In the label of the billboard, a short text described the lesson. The children chose one of the billboards and entered a corresponding pedagogical stage. Once a lesson was completely visualized, the colour of the billboard changed.

A detailed presentation of the use of the application would follow the subsequent steps:

Lesson1/Level 1 ("the visual story of the place"):

The AR agent augments the image of the real context with a reconstruction of a 3D Roman *villa rustica* seen from the front; when the user clicks on this object, a film with the re-enactment of a technology starts. After the film ends the user can return to the real image of the context and select another lesson.

Lesson 2/Level 2 ("the visual story of objects") is composed of several stories.

Level 2a ("the visual story of objects—the loom"): The AR agent provides a virtual tour of the Roman villa and stops the tour in front of the 3D reconstruction of a loom. By clicking on this image, a 2D image of the loom and a series of texts with historic and technical explanations appear followed by another 3D reconstruction, which can be rotated to see all the details. Once the film ends, the user can return to the real image of the context and choose another lesson.

Level 2b ("the visual story of objects—furniture"): The AR agent provides a virtual tour of the Roman villa, and stops the tour in front of the masters' room, equipped with furniture and daily objects. By clicking on this image, a 2D image of the room appears followed by another 3D reconstruction of the room, which can be rotated to see all the details. Once the film ends, the user can return to the real image of the context and choose another lesson.

Lesson 3/Level 3 ("the technological digital storytelling"):

The AR agent completes the real context with the reconstruction of a Roman *villa rustica* seen from the front; by clicking on different objects, a series of video films on technological gestures are opened, which will help the user to better understand the process studied. During each learning lesson the users can send an e-mail message with comments or an image capture using a Facebook or Twitter account.

Lesson 4/Level 4 ("the craft approach"):

This level is the applied for the practical part of the hyper-story. For example, the teaching of the weaving techniques to children between the ages of 8 and 12 years old was intended to develop the coordination of the gestures, a high degree of skill, the attention and, last but not least, the endurance. Although it is intended to be a solitary activity, the teaching of weaving to groups of children could develop their interactivity and cooperation skills in solving technical problems. During the practical workshops a skilled group of children was identified, and the contact with them was continued after the completion of the experiments, with the help of Skype and Facebook. The technological lessons were completed with information on the history of textiles, as well as with the presentation of unconventional techniques and new approaches to fiber art during the videoconferences which followed the experimental campaign in the summer of 2012.

The children rapidly understood how to use performant devices but had to repeatedly try the application in order to understand our learning scenario. They provided us useful suggestions regarding the design of the application (e.g. UI interface, text messages), in order to be motivated to continue to use the application beyond the experimental phase.

3.6 Educational Novelty and Outcomes

In our project we have experimented a "learning-by-playing" teaching method and mobile-learning in real and virtual contexts, as a personal style of learning. Also some components of social learning were integrated.

The learning application implements hypermedia narratives with AR and several software agents, which communicate with each other in order to create an adaptive and coherent application that serves our teaching strategy: the learner has the freedom to choose a learning path from three learning levels; the overall learning process is guided to help the learner to obtain a coherent knowledge.

The customization of the application according to the user profile also captures the children's interest. This was confirmed by a survey which asked the children and teachers about their preference between an application insensitive, or sensitive, to the user profile.

The application works in a geographical context, which conditions the children to place themselves outdoor and seek the historical areas, geographically defined.

Positioning in the real context stimulates an active learning through a play-like discovery. The novelty of the project approach resides in the simultaneous presentation of information on the objects or technologies, and that of the culturally formative context.

We tested the application on small groups of children. These were presented with the whole application and were then given the mobile devices to use the application in several campaigns lasting 2–3 weeks. The last such campaign took place in April 2013. This approach was based on the fact that the learning process is slow. We periodically verified, together with the school principal and the fiber artists, the changes that occurred in the children's levels of knowledge and skills.

As observed during the experiments, the mobile Augmented Reality application for educational purpose had a proven, positive impact on children and young audiences, both in formal and informal educational settings. This is due to the contextual, user-centric information and the direct connection with true learning contexts, in this case, both real-life surroundings and virtual reconstructions. These types of applications are examples of blended learning solutions, which in conjunction with traditional methods can be efficient learning tools. Our learning paradigm helped both children and teachers, the latter gaining a better understanding of where to insist on teaching and evaluating the learning outcome.

We also evaluated these changes during the practical workshops ("the craft approach") with a group of school children of the Vădastra School. Following the first test period, the school children proved that they had appropriated the specific terminology: loom frame, warp, weft, sheds, heddle rod. Also they had learned the difference between a plain weave, a kilim weave and a knotted weave and different types of weaver's knots. The problems encountered (not only in this age group but in any age group) were noticing and understanding the difference between sheds and picking up the right threads using the hand, keeping a straight border. Corrections were made through successive demonstrations accompanied by explanations. Following the first workshop a group of school children that had the skill and wish to master weaving techniques was identified.

The next experiments introduced the children to different weaving technologies (prehistoric, Roman and modern). What followed was a tour of the experimenting site, a tour in which the pupils learned and worked with each technology. In the prehistoric house, during a series of demonstrations, they displayed their understanding of the mechanism of weaving and various weaving stages. In explaining the techniques and tools the teacher had used information already acquired in the previous experiment. The pupils discovered that the few major differences between the two bar loom, with which they had worked before, and the backstrap loom were: the tension system using the weaver's body and a fixed point, the maximum width of the fabric, the presence of heddle rods, and auxiliary tools such as the weaver's sword.

The educational exercises carried out with the children from the Vădastra village school demonstrated that this method of "teaching in virtual contexts" had a strong influence on the young generations. How this method can represent a successful means of preserving part of the identity of traditional societies in the third millennium will only be fully appreciated over a longer follow-up period, but we already have obtained positive results.

3.7 Conclusions and Future Work

The teaching experiments that we conducted during the course of one year, both by traditional methods and with an educational application, allowed us to experience a paradigm with original pedagogical elements, designed for learning traditional technologies in their historical context. We believe that expanding the perception of the physical world with AR virtual elements can help cognitive and educational processes.

In this chapter we presented significant results of our research regarding the implementation of our MapsofTime-LearningTool application, focusing on the role of the software agents for creating narrative e-learning tools and on evaluation of the effective educational outcomes. We conceptualized a learning paradigm and applied it using advanced software technologies, i.e. software agents and Augmented Reality on Android mobile devices. As much as possible, open and flexible software technologies were used. The prototype application has the architecture and functionalities of a native application for Google Android mobile platform, which combines different metaphors and technologies: geographic Points of Interest (Geo-POIs) for discovering and accessing different learning levels; digital stories based on inter-connected multimedia content (text, graphic, video); software agents and social media.

The navigation through the multimedia space of a learning level is performed by simple interaction with virtual objects in the AR view. The overall learning process is supervised by a learning agent. The software agents' tasks and the agent-based environment controlling the communication among them were programmed using the JaCa-Android framework, which provides a high level of abstraction and facilitates the application development. These agents have different roles within our application: monitoring and storage of the completion status of each learning level for each registered user; learners' assistance in using the lessons in a consistent manner, based on the history of the application usage; achievement of a general application autonomy.

The AR application was designed to create a mixed reality environment, where what mattered was the user's positioning in the defined geographic area, and not his ability to locate a particular geographical point. The AR application by itself could not produce important learning outcomes. That is why the integration of software agents in an Augmented Reality application for the implementation of complex educational applications is justified as the integrated AI control can actively engage and support children in the learning process and provides a coherent pedagogical structure. This kind of application can support the blended learning paradigm, which is based on finding learning environments and media complementary to the traditional classroom ones. The agent-based paradigm is also very well suited to a non-linear learning style, which has recognized cognitive outcomes.

The learning paradigm and the digital narratives are original, made of a combination of technical video lessons with artistic performance videos, object annotations and simple animations. The application and multimedia lessons will not by themselves make the children achieve the level of skill required to manufacture traditional objects, this necessitates years of practice, but they can make them understand how that object was produced in past times. Our modern learning paradigm is intended to be used together with traditional face-to-face learning sessions.

As future developments we will extend the application to other traditional technologies (ceramics, glassware) and historic periods (e.g. prehistory). The agent functionalities will be further developed with predictive functions, e.g. to provide further recommendations based on user learning preferences. An Android SQLite database will be employed to store historic usage data, to be further analyzed by means of Business Intelligence (BI) tools in order to offer valuable information to the school community. These data could also help improve the application and prototype certain functions in order to promote the adoption of these kinds of e-learning solutions.

The user base will be extended by testing the application in other schools.

Acknowledgments We wish to thank to the editor Professor Mirjana Ivanovic for her moral support during the elaboration of the text, Mrs. Laura Voicu, the principal of the Vădastra school, Vădastra village, Olt county, for the help during the summer campaigns, Professor Andreea Hasnaş, Professor Viorica Slădescu and technician Elena Haut (NUA) for their collaboration and to the art students and children who volunteered in the project. Last but not least, our gratitude goes to Mr. Bogdan Căpruciu for reviewing the translation.

The project was financed by an exploratory research grant PN II IDEI (The Maps of Time. Real communities, virtual world, experimented pasts, Director Prof. Dragoş Gheorghiu).

Images by D. Gheorghiu and A. Şerbănescu.

References

1. Alsheail, A.: Teaching English as a second/foreign language in a ubiquitous learning environment: A guide for ESL/EFL instructors (2010)
2. Azuma, R.: A survey of augmented reality. Presence Teleoperators Virtual Environ. 6(4), 355–388 (1997)
3. Balcisoy, S., Kallmann, M., Torre, R., Fua, P., Thalmann, D.: Interaction techniques with virtual humans in mixed environments. In: Proceedings of International Symposium on Mixed Reality (ISMAR 2001), Tokyo, Japan (2001)
4. Barakonyi, I., Weilguny, M., Psik, T., Schmalstieg, D.: Monkey bridge: Autonomous agents in augmented reality games. Adv. Comput. Entertainment Technol., 172–175 (2005)
5. Barber, E.J.W.: The Development of Cloth in the Neolithic and Bronze Ages with Special Reference to the Aegean. Princeton University Press, Princeton (1992)
6. Broudy, E.: The Book of Looms: A History of the Handloom from Ancient Times to the Present. Hannover and London (1979)

7. Cassell, J., Ananny, M., Basu, A., Bickmore, T., Chong, P., Mellis, D., Ryokai, K., Smith, J., Vilhjálmsson, H., Yan, H.: Shared reality: Physical collaboration with a virtual peer. In: Proceedings of Human Factors in Computing System (CHI 2000), pp. 259–260. The Hague, The Netherlands (2000)

8. Cavazza, M., Charles, F., Mead, S.J., Martin, O., Marichal, X., Nandi, A.: Multimodal acting in mixed reality interactive storytelling. IEEE Multimedia 11(3), 30–39 (2004)

9. Clark, R.: Media will never influence learning. Educ. Technol. Res. Dev. 42(2), 21–29 (1994)

10. Conrad, R., Donaldson, A.: Engaging the Online Learner: Activities and Resources for Creative Instruction. Jossey-Bass, San Francisco (2004)

11. Costanza, E., Kunz, A., Fjeld, M.: Mixed reality: A survey. In: Invited Book Chapter in Human Machine Interaction, LNCS 5440, pp. 47–68. Springer, Berlin (2009)

12. Dede, C.: Immersive interfaces for engagement and learning. Harvard Graduate School of Education (2009)

13. Dobrincu, D., Iordachi, C. (eds.): Taranimea si puterea. Procesul de colectivizare a agriculturii in Romania (1949–1962). Iasi, Polirom (2005)

14. Duffy, T.M., Jonassen, D.H.: Constructivism: New implications for instructional technology. In: Constructivism Technology of Instruction, pp. 1–16. Routledge, London (1992)

15. Geddes, S.J.: Mobile learning in the 21st century: Benefit for learners. Knowl. Tree e-journal: An ejournal Flex. Learn. VET 30(3), 214–228 (2004)

16. Gheorghiu, D.: Le projet Vădastra. Prehistorie Europeenne 16–17, Liège (2001)

17. Hildebrand, A., Daehne, P., Christou, I.T., Demiris, A., Diorinos, M., Ioannidis, N., Almeida, L., Weidenhausen, J.: Archeoguide: An augmented reality based system for personalized tours in cultural heritage sites. In: International Symposium of Augmented Reality 2000 (2000)

18. Hoffman, S.: Elaboration theory and hypermedia: Is there a link? Educ. Technol. 37(1), 57–64 (1997)

19. Holza, T., Campbella, A.G., O'Harea, G.M.P., Stafford, J.W., Martin, A., Dragonea, M.: MiRA—mixed reality agents. Int. J. Hum. Comput. Stud. 69(4), 251–268 (2011)

20. http://www.amire.net

21. http://jade.tilab.com

22. http://jaca-android.sourceforge.net

23. http://www.activecomponents.org

24. http://www.agentfactory.com/index.php/Main_Page

25. http://www.cs.uu.nl/3apl

26. http://jason.sourceforge.net

27. http://cartago.sourceforge.net

28. https://developers.facebook.com/docs/facebook-login

29. http://www.metaio.com/products/creator

30. http://www.layar.com

31. http://www.wikitude.com

32. http://dev.junaio.com

33. Jung, Y.: Building Blocks for Virtual Learning Environments. Fraunhofer IGD, Darmstadt, WSCG2008 (2008)

34. Jung, Y., Behr, J., Graf, H.: X3DOM as carrier of the virtual heritage (2011)

35. Laurel, B.: Interface agents: Metaphors with character. In: Bradshaw, J.M. (ed.) Software Agents, pp. 67–77. MIT Press, Cambridge (1997)

36. Lockard, J., Abrams, P.D., Many, W.A.: Microcomputers for Twenty-first Century Educators, 4th edn, p. 219. Longman, New York (1997)

37. Maes, P., Darrell, T., Blumberg, B., Pentland, A.: The ALIVE system: Wireless full-body interaction with autonomous agents. ACM Multimedia Syst. 5(2), 105–112 (1997)

38. Nwana, H.S.: Software agents: An overview. Knowl. Eng. Rev. 11(3), 205–244 (1996)

39. Opresco, G.: L'Art du paysan roumain. Roto-Sadag (1943)

40. Papagiannakis, G., et al.: VR & AR Frameworks for Virtual Character Simulation State-of-the-art. Brussels, AR Concertation Meeting (2002)

41. Perelman, L.J.: School's out: A Radical New Formula for the Revitalization of America's Educational System. Avon Books, New York (1992)
42. Richey, R.C.: The Legacy of Robert M. Gagné. ERIC Clearinghouse on Information and Technology, Syracuse, NY (2000)
43. Sánchez, J., Lumbreras, M., Bibbo, L.M.: Hyper stories for learning. In: Proceedings of International Workshop on Hypermedia Design, pp. 239–248. Montpelier (1995)
44. Sanchez, J., Lumbreras, M.: Hyper stories as a metaphor for educational software. Retrieved http://users.dcc.uchile.cl/~jsanchez/Pages/papers/metaphorforeducation.pdf (2009)
45. Santi, A., Guidi, M., Ricci, A.: Exploiting Agent-Oriented Programming for Developing Android Applications. University of Bologna, Bologna (2011)
46. Santi, A., Ricci, A.: A programming paradigm based on agent-oriented abstractions. Int. J. Adv. Softw. 5(1–2) (2012) http://www.iariajournals.org/software/
47. Sârbu, C., Gheorghiu, D.: From fragments to contexts: Teaching prehistory to village children in Romania. In: An Interdisciplinary Perspective Teaching Children About the Past, pp. 334–346 (2007)
48. Schlosser, C., Burmeister, M.: The best of both worlds. Tech. Trends 43(5), 45–48 (1999)
49. Schulmeister, R.: Hypermedia Learning Systems. Hamburg (2012)
50. Shoham, Y.: Agent-oriented programming. Artif. Intell. 60(1), 51–92 (1993)
51. Smith, R.G.: The contract net protocol: High-level communication and control in a distributed problem solver. IEEE Trans. Comput. C29(12) (1980)
52. Stahl, H.: Sociologia Satului Romanesc Devalmas Romanesc. Bucharest, Fundatia Regele Mihai I (1946)
53. Steinmetz, R.: Multimedia-Technologie: Einführung und Grundlagen. Springer, Berlin (1993)
54. Wild, J.P.: The Roman horizontal loom. Am. J. Archaeol. 91, 459–472 (1987)
55. Wooldridge, M., Jennings N.R.: Intelligent agents: Theory and practice. Knowl. Eng. Rev. (1994)

Chapter 4
Inter-university Virtual Learning Environment

**Andrej Tibaut, Danijel Rebolj, Karsten Menzel
and Ricardo Jardim-Goncalves**

Abstract Virtual learning environments are irreversibly shaping the way of teaching and learning at universities. They have become more and more popular as practitioners recognized the advantages virtual learning environments offer when applied in online teaching and learning. Nevertheless, most of the virtual learning environments do not provide straightforward support for inter-university interoperability between heterogeneous learning environments. This paper presents a use case of ITC-Euromaster with collaboration concept for inter-university interoperability. Furthermore, a taxonomy, an ontology and a high level architecture using intelligent software agents that enable dynamic inter-university interoperability are proposed to further develop an existing virtual learning environment used among four European universities.

Keywords Inter-university · Virtual learning environment · Federated approach · High level architecture · Dynamic interoperability · Ontology · Software agents

A. Tibaut (✉) · D. Rebolj
Faculty of Civil Engineering, University of Maribor, Maribor, Slovenia
e-mail: andrej.tibaut@um.si

D. Rebolj
e-mail: danijel.rebolj@um.si

K. Menzel
University College Cork, Cork, Ireland
e-mail: k.menzel@ucc.ie

R. Jardim-Goncalves
New University of Lisbon, Lisbon, Portugal
e-mail: rg@uninova.pt

M. Ivanović and L. C. Jain (eds.), *E-Learning Paradigms and Applications*,
Studies in Computational Intelligence 528, DOI: 10.1007/978-3-642-41965-2_4,
© Springer-Verlag Berlin Heidelberg 2014

4.1 Introduction

Virtual learning environments (VLEs) in higher education are typically used for facilitation of online teaching and learning (OTL). Online teaching and learning occurs thru online participation (OP) within different organizational forms like courses, labs, meetings, etc. Facilitation mechanisms (FM) that enable online participation are software and hardware systems like virtual classrooms (VC) and learning management systems (LMSs). The above-abbreviated VLE concepts are involved and applied to educational process at modern universities. However, application of VLE to educational process is not as straightforward as expected because of the two apparently contradictory assumptions: a rather pessimistic assumption is (a) that there is no invention in online teaching and learning when compared to the proven face-to-face concept that works well in traditional classroom and another, rather enthusiastic assumption is (b) that well designed course and proper technology is simply enough for successful teaching and learning online. Of course, both assumptions are false. After while pessimism and enthusiasm fade and teachers are looking for assessment methodologies for OTL that would lead to improvement of their online activities. On the other hand understanding of technical dynamics of OTL, its requirements and improvements towards inter-university enterprise systems' interoperability best describe the aim of the chapter. Our VLE of interest is the ITC Euromaster (ITCEM—European Master Programme in Information Technology in Construction). The ITCEM [1–3] is an e-learning award-winning [4] inter-university virtual learning community based on common course pool. The inter-university concept involves assumption that VLE has undeniable strengths for integrating fragmented expert domain knowledge that exists at different universities. In our case this is the knowledge about IT in civil engineering (also IT in construction or construction informatics). An inter-university VLE requires also a shared understanding of existing organizational structures: a common taxonomy for information classification and organization of a subject that is common for all members of the virtual learning community. The course pool is a unique mechanism for sharing courses between providers and consumers. Course pool provider is a partner institution providing a teacher, who contributes a unit of study to the pool. Course pool consumer is a partner institution that uses a course from the course pool in its curriculum and for its students.

The chapter is structured in five sections: after the introductory section, the next section presents related work about inter-university collaboration concepts and continues with taxonomy that provides basic understanding about the domain, inter-university VLE. In the third section the realized VLE, namely ITC Euromaster, is described. Section four describes concept for implementation of dynamic interoperability between university information systems belonging to participating universities in the ITC VLE. The implementation follows the federated architecture pattern. Section five describes challenges for further research and development of the ITC Euromaster platform. Section six concludes the chapter.

4.2 Review of Related Work

The purpose of this section is to review the key collaboration concepts that contribute to knowledge sharing at different inter-university levels. This way, we aim at improving our understanding of different channels of collaboration between universities in educational process within VLE. In the second part of the section taxonomy is developed, which provides an agreed vocabulary for the inter-university VLE. The taxonomy also aids in lowering cost of misunderstanding because of different vocabularies used at participating universities. Last part of the section reviews advances in enterprise interoperability patterns and technologies suitable for development of seamless inter-university interoperability for ITCEM.

4.2.1 Inter-university Collaboration Concepts

Knowledge sharing through collaboration between universities is a type of inter-organizational knowledge transfer through which one organization learns from the experience of another [5]. Inter-university knowledge sharing is facilitated through local, regional and global collaboration mechanisms. Local and regional inter-university collaboration typically runs within the framework of selected national universities [6, 7]. Globalization in education (i.e. Bologna process) leads to internationalization [8] and brings concepts like transnational, borderless and cross-border education [9] for better mutual and intercultural understanding and learning and better mobility of academic staff and students. The concept of inter-university knowledge transfer in education mostly takes the form of dual degree programmes. Dual degree programmes allow students to obtain degrees from both participating universities for a single programme of study. In [10] authors described a transnational, dual-degree programme in the context of knowledge transfer process. They developed a theoretical framework (Fig. 4.1) for inter-university knowledge transfer process.

Alternatives to the controlled dual-degree programmes are more flexible inter-university programmes where collaboration is at the level of units of study (i.e. joint courses). Such collaboration brings together students from at least two universities to work as a team. As the main contribution and benefit of the collaboration their implementers emphasize the benefit of sharing units of study which are not offered by students' own university [6]. It is also important that the enrollment process for students is transparent, so that students can enroll into their chosen units of study in the usual form at their home university [7]. In such collaboration communities special attention needs to be paid to a shared and synchronized timetable between universities, shared staff and courses such as to avoid redundancy.

Multidisciplinary inter-school programmes offer creation of unique degrees. Such approach enables custom design of a degree programme (multiple-area

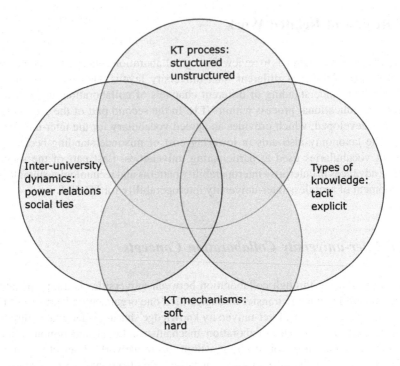

Fig. 4.1 Inter-university knowledge transfer conceptual framework

programmes, thematic programmes) including units of study and disciplines from across a single university [11]. The concept is known in Continuous learning and Life-long learning programmes.

Joint degree programmes are established in cases where a single university cannot cover a specific field of knowledge. In such cases, collaboration between universities is a way to complement the dispersed knowledge where former competitive programmes at different (national) universities were able to overcome their mutual suspicion and have developed together a joint programme [12].

A decoupled version of the joint degree programmes is where involved universities decided not to run the whole programme but to incorporate some of the units of study of the programme into the existing curricula. The ITC Euromaster programme uses the concept of a course pool to facilitate collaboration between universities [3]. Any partner institution can include any number of existing units of study (courses) in its own curricula.

From the review the following collaboration models can be identified:

- Shared programme: this is rather traditional form of collaboration (by design) where universities intentionally design their curriculum to increase their collaborative capacity. The collaboration is best designed through the curriculum mapping process [13]. Results of the process are tightly coupled dual-degree programmes (level of collaboration is "programme of study") of equal standard.

Such collaboration results in many potential students abandoning their plans to move to partner university [10] which increases number of students at home university and decreases individual study costs. The collaboration is based on moving staff-pool and/or support of virtual learning environment.

- Shared course pool: this is a loosely coupled concept of inter-university collaboration, which is a mediation workspace between course providers and course consumers. Course consumers use the course pool mechanism to obtain requested units of study, while course providers maintain and deliver units of study to the pool [3].
- Shared teacher pool: this is an old but interesting concept for providing substitute teachers in large school districts. In [14] authors suggested an optimal size of permanent pool of substitute teachers as one possible method for improving substitute teacher quality as well as reducing school cost.

Concluding, we hypothesize that the shared course pool is the most flexible and open collaboration model for an inter-university virtual learning community.

4.2.2 Taxonomy for Inter-university Interoperability

Practical knowledge sharing through collaboration in virtual learning environment at university level [15] is influenced by research that ranges from virtual networks [16], virtual teams [17–19], professional virtual communities of practice [20–24], virtual enterprises [25], agent-based technologies for efficient knowledge sharing and personalization of e-learning services [26–32] and semantic web technologies [33, 34].

At a university collaborative knowledge sharing supported by e-learning may occur at different granularity levels within the generic nine-level taxonomy for inter-university collaboration (Fig. 4.2): sharing between universities (inter-university), schools (faculty, department, college), degree schemes (graduating curriculum, degree, sub-degree), programmes of study, tracks of study (specializations), modules of study (specialized sub-tracks), units of study (courses, classes or subjects), course units (lessons, activities) and/or at the level of individual learning objects (lesson chapters).

Shared understanding of existing organizational structures at a university is needed for new and innovative collaboration models, new delivery modes, new policies, new assessment models, new funding models, and new skills on the part of the students, faculty members and administrators, which eventually may also lead to new virtual universities [35–37]. While on the other hand effective digital knowledge sharing at university level education leads to higher Bloom's levels of conceptual understanding [38, 39].

The nine-level taxonomy is a classification scheme for the categorization of organizational structure at universities. Common taxonomy and its methods are useful for information classification and organization of a subject that is common

Fig. 4.2 Basic taxonomy for inter-university collaboration

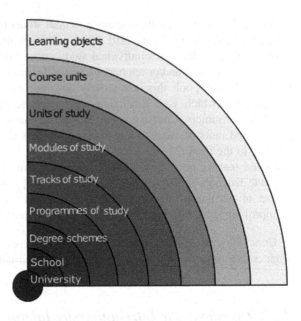

for all members of the virtual learning community. Specifically, it defines controlled vocabulary that helps overcome differences of language usage in different members (individuals or organizations) of the virtual learning environment. Semantic technologies complement taxonomy methods. Implementation of semantic technologies can also include ontologies. An ontology identifies, captures and arranges key concepts in a hierarchy that is often referred to as structural knowledge taxonomy [40].

Referring back to the previous section and applying the taxonomy to the course pool we can conclude that the course pool facilitates knowledge transfer at the level of units of study (courses, classes or subjects). While on the other hand the tightly coupled dual-degree programmes facilitate knowledge transfer at the level of programmes of study (inter-programme).

4.2.3 Interoperability Technologies

In the past, universities competed for students against each other. Nowadays, modern universities must collaborate in order to be more competitive. Universities that adapt to collaborative online environments broaden their enrolment base towards international students. Consequently, universities can be regarded as enterprises that interoperate with many different heterogeneous enterprises implementing different business processes, technologies and semantics. Generally, "to inter-operate" implies that one system performs an operation on behalf of (or for) another system. Enterprise Interoperability (EI) is well-established applied systems' research area, studying the problems related to the concerns and barriers of systems', applications' and data interoperability [41]. According to the INTE-ROP Enterprise interoperability Framework (now CEN/ISO 11354 standard) various dimensions of EI are:

- Interoperability concerns: levels within an enterprise at which the interoperation occurs (data, service, process, business).
- Interoperability barriers: obstacles to interoperability (conceptual, technological, and organizational).
- Interoperability approaches: approaches for elimination of interoperability barriers (integrated, unified, federated).

The objective of the INTEROP framework is to tackle interoperability problems through the identification of barriers that prevent interoperability to occur. The first two dimensions (interoperability concerns and barriers) constitute the problem space of enterprise interoperability. After identifying the problem space, the "interoperability approaches" propose solutions to this problem space:

- **Integrated approach** means that there exists a common format for all models. Diverse models are built and interpreted using/against the common template. This format must be as detailed as the models themselves. The common format is not necessarily an international standard but must be agreed by all parties to elaborate models and build information systems.
- **Unified approach** means there is a common format but it only exists at meta-level. This format is not an executable entity as it is the case in integrated approach. Instead it provides a mean for semantic equivalence to allow mapping between models and applications. Using the meta-model a translation between the constituent models is possible even though they might encounter loss of some semantics or information.
- **Federated approach** aims to establish the interoperability on the fly, which means that the adoption of the approach should not impose the existing models, languages and methods of work as the integrated approach.

The integrated and unified approaches have been well researched and implemented, but the federated approach is still an ongoing research. In the revised Enterprise interoperability research roadmap (European Commission, [42]), development of federated approach for interoperability is considered as one of the grand challenges. Today, most of existing interoperability solutions propose integrated or unified approaches, which are not satisfactory to dynamic networking enterprises.

Practical requirements of inter-university VLEs from the research and development in the EI domain should be that partner universities participating in a VLE don't need to think about re-engineering of their legacy systems or building up an integrated platform and proposing novel methods and frameworks to support collaboration. Instead, participating universities' information systems should use technology that enables establishing interoperability on the fly.

Technologies that are helpful for the development of federated EI are:

- **Model driven architecture** (MDA): this methodology has been defined and adopted by the Object Management Group (OMG) in 2001 and updated in 2003. It is designed to promote the use of models and their transformations to consider and implement different systems. It is based on an architecture defining four levels, which go from general considerations (CIM—Computation Independent level and PIM—Platform Independent Model) to specific ones (PSM—Platform Specific Model and Coding level).
- **Model driven interoperability** (MDI): the approach considers interoperability problems at the enterprise model level instead of only at the coding level. The main goal of MDI, based on model transformation, is to allow a complete follow-up from the expression of requirements to the coding of solutions and also to provide a greater flexibility thanks to the automation of these transformations. MDI concepts were developed in the Task Group 2 (TG2) of INTEROP-NoE [43].
- **Reverse engineering model**: after MDA, OMG launched another research activity leading to what was later called Architecture Driven Modernization (ADM) [44] Reversing the MDA lifecycle, ADM is discovering models from the coding level of legacy information system, such as UML models, Knowledge Discovery Meta-model (KDM) and Abstract Syntax Tree Meta-model (ASTM) [45]. KDM and ASTM are aimed to satisfy someone interested in discovering more specific models from a legacy system.
- **High level architecture** (HLA): the HLA Evolved 1516 [46] is a software architecture specification that defines how to create a global software execution composed of distributed simulations and software applications. The Defense Modeling and Simulation Office (DMSO) of the US Department of Defense (DOD) originally introduced this standard. The original goal was reuse and interoperability of military applications, simulations and sensors. In HLA, every participating application is called a "federate". A federate interacts with other federates within a HLA federation, which is in fact a group of federates. The interface specification of HLA describes how to communicate within the federation through the implementation of run time infrastructure (RTI). Federates interact using services proposed by the RTI. They can notably "Publish" to inform about an intention to send information to the federation and "Subscribe" to reflect some information created and updated by other federates. The information exchanged in HLA is represented in the form of classical object class oriented programming. The two kinds of object exchanged in HLA are class Object and class Interaction. Class Object contains object-oriented data shared in the federation that persists during the run time; data from class Interaction are information sent and received between federates. These objects are implemented in XML format. HLA also supports web services for easier interaction with legacy systems.

4.3 Use Case: ITC Euromaster

ITC-Euromaster (European Master Programme in Information Technology in Construction—ITCEM) is one of the most unique and award winning e-learning paradigm [3, 4] in higher education offered to civil engineering bachelors. The ITCEM has developed an e-learning model of education based on the course pool and participation of academic institutions being course providers and/or course consumers. Specialized courses, offered by five European Universities (University of Maribor from Slovenia—UM, University College Cork from Ireland—UCC, Dublin Institute of Technology (CITA) from Ireland—DIT, University of Ljubljana from Slovenia—UL and University of Technology Graz from Austria—TUG), cover a wide area of IT in Construction. Graduates from participating universities represent a unique profile of an ITC specialized Civil Engineer that is increasingly recognized and demanded by the Architecture/Engineering/Construction (AEC) sector.

The courses (units of study) of the ITCEM Course Pool were initially developed by academics from nine European universities as part of a EU-funded Socrates Erasmus project between 2002 and 2005. The main purpose of the project was to develop a curriculum on IT in Construction to give students the possibility to extend their knowledge in research, development, and application of computer and information science in civil and building engineering. The result in 2005 was a European Master's curriculum in Construction IT, complements the existing portfolio of teaching programs and should meet the growing demand for such skills. In the case of the institutions already offering ITC courses, the project is providing the added value of a European dimension for their existing programmes.

After the Socrates Erasmus project was finished in 2005 only two of the initial partner universities, namely University of Maribor, Faculty of Civil Engineering and University College Cork, Department of Civil and Environmental Engineering were successful with the incorporation of the ITCEM-like curriculum into their existing curriculums. That practically means that any partner institution can include (consume) any number of existing units of study (courses) from the ITCEM Course Pool into their own curricula. Today we know that the lowest common denominator for ITCEM virtual learning community is a shared course pool along with a shared and synchronized timetable for the courses to ensure one-time delivery for all course consumers.

The diagram on Fig. 4.3 shows the most common AS–IS Use cases within a university information system (UIS). It is important to note that the ITCEM VLE depends on the data distribution from the UIS. The data is semi-manually exported from UIS and imported to ITCEM VLE. In the following sections a concept of the new architecture with higher interoperability level is presented.

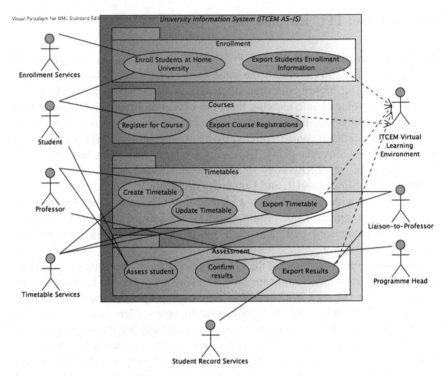

Fig. 4.3 AS–IS use case diagram for UIS in relation to ITCEM

4.3.1 The ITCEM Course Pool

Currently, ITC-Euromaster (single entry point is http://euromaster.itcedu.net) delivers over 10 courses developed and offered by academics from 5 European Universities in Slovenia, Ireland and Austria. The courses form an ITCEM Course Pool (mostly delivered courses are: The Role of Construction Informatics, eBusiness in Construction, Automation in Construction, Computer Mediated Communication, Applied Knowledge Management, Software Engineering, Interoperability and BIM, Computer Aided Facilities Management). Course pool provider is a partner institution/author who contributes one or more units of study (courses) to the pool. A course can have more than one provider from different partner institutions (i.e. joint delivery of a course). Any institution/author (provider) with the knowledge in the field of ITC is welcomed. Once accepted by the steering committee, the new course is included in the pool. Course pool consumer is a partner institution that uses one or more courses from the course pool in its curriculum. Course consumers from partner institutions are students who enroll for the course at their home institution. However, the students participate in a single virtual remote class, which is delivered by its provider.

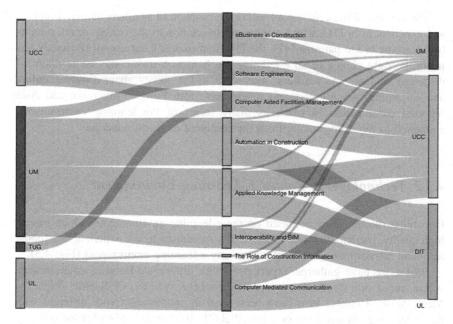

Fig. 4.4 The concept of the course pool; sharing of units of study (courses)

The initial course pool content (between 2005 and 2009) consisted of 12 courses: The role of construction informatics, Data structuring and databases, Information modeling and retrieval, Modeling and visualization, Software engineering, Knowledge management, Engineering Artificial Intelligence, Computer mediated communication, Mobile computing in construction, Computer integrated construction, Virtual enterprises and eBusiness. Since then the course pool content was distilled and now contains 8 core courses plus some other (electives at partner institutions).

Responsibility for the course content, namely course units and learning objects in digital form, has course provider.

The diagram on the Fig. 4.4 shows universities as course providers (left) and universities as course consumers (right) in relation (provided by and consumed by) to the courses from the pool (middle). In the year 2012/2013 four universities (UM, UCC, UL, TUG) are course providers and 4 universities (UM, UCC, DIT, UL) are course consumers.

The ITCEM course pool content is focused towards students who have finished their undergraduate studies with a university degree in civil, building or structural engineering as well as architecture. At partner institutions (consumers) the courses are used as part of different forms of Master's degree curriculum in Construction Information Technology. This shall enable students to continue with the relevant PhD study or immediately start to work in the industries with a specific knowledge of IT. The need for such new profile has already been recognized.

The concept of the course pool solves the problem of adequate human resources and experiences in ITC, which are scarce. In relation to the current developments of e-learning usage patterns [47], the ITC-Euromaster Course pool obviously has made the right choice of an e-learning mechanism to overcome the problem of dispersed students and teachers. According to the survey about distance, online and e-learning, practitioners and researchers from Europe, Australia, and Asia reported participation in twice as many forms of the learning environments compared to those originating from the continent of North America.

4.3.2 Technology: A Robust E-Learning Environment

E-learning models must be based on the e-learning environments, which ensure high quality, participation and productivity. According to our experiences a robust technical infrastructure is a vital part of any e-learning environment (Fig. 4.5).

So far we have gathered experiences with different e-learning environments where audio or videoconferencing (HorizonLive, VCON, CUSeeMe, ClickTo-Meet, Adobe Connect Pro) have been used in combination with different web based learning management systems (WebCT, Blackboard, Moodle) for delivery of courses from the ITCEM course pool. Our experiences, enriched with those of many researchers in the field of using IT for education, led to ideas of an ideal environment to effectively support open distance teaching and learning.

The current ITC Euromaster e-learning environment consists of three components: a portal, the course management system Moodle together with file repository in DropBox and a Virtual Classroom facilitated by Adobe Connect Pro. The main function of the first is to enable access to teaching and learning material repository as well as other relevant functions (e.g. forums) and information (e.g. teacher and student list, timetables, grade book, etc.). The learning management system

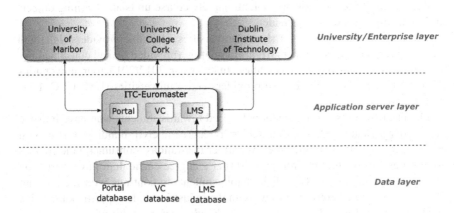

Fig. 4.5 ITC-Euromaster system architecture

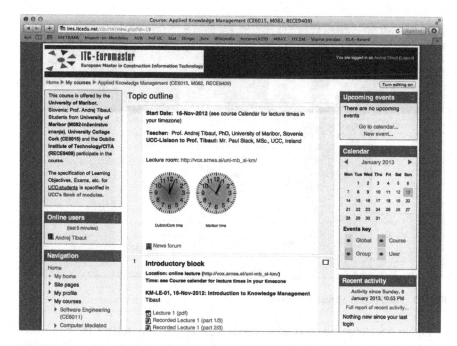

Fig. 4.6 The ITCEM learning management system

(Fig. 4.6) is based on the Moodle—Modular Object-Oriented Dynamic Learning Environment (also categorized as a CMS—Content Management System or VLE—Virtual Learning Environment). The Virtual classroom [1, 15, 48] enables teachers synchronous communication with their students. A participant list, chat, audio and video control, web links, document sharing, application sharing and a whiteboard are the basic parts of the virtual classroom (Fig. 4.7). ITCEM community has access to three Virtual classrooms. They are geographically distributed (one in Slovenia, two in Ireland) installations of Adobe Connect Pro.

4.4 Specification of the Federated Inter-university Virtual Learning Environment

This section presents a specification for development of a framework based on the previously described interoperability approach called federated approach. The framework is HLA compliant, which means that interoperability between participating university information systems (UIS) in the virtual learning environment ITCEM can be setup rapidly, on the fly. The notion "on the fly" is important for framework scalability when new partners will join the current consortium of partner universities. The framework also introduces use of ontology during communication between UISs.

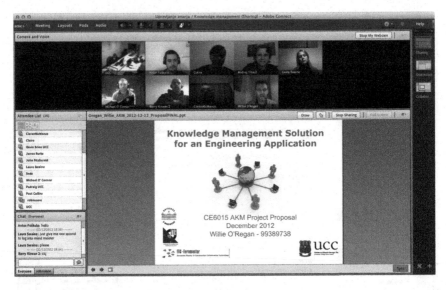

Fig. 4.7 The ITCEM virtual classroom

Figure 4.8 shows new ITCEM system architecture that utilizes the new HLA based framework. Existing literature suggests gradual development (evolvement) of HLA based architectures [49]. The modernization process requires following steps:

- Describing the information extracted out of the artifacts of existing information system: Open source software MoDisco [50] provides generic tools for static analysis of existing systems (i.e. University information system). The tools can generate complex models (meta-models) out of existing systems. Model Discover (MoDisco) is an Eclipse Generative Modeling Technologies (GMT)

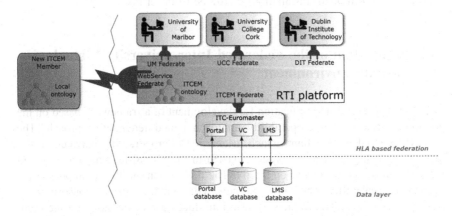

Fig. 4.8 The new ITCEM system architecture based on HLA—high level architecture

component for model-driven reverse engineering. MoDisco can extract XML model, KDM model, KDM code model, JAVA code model, UML model, etc. from existing legacy systems (i.e. university information system). The complex models, called meta-models, emulate (modeling world) the existing enterprise information systems (real world, i.e. university information system). The idea is to automate the process of meta-model creation, which in MoDisco is possible with the creation of "discoverers". The discoverers extract necessary information from the system (i.e. databases, Java and C++ project files, etc.) in order to build a model conforming to the previously defined meta-model. The way to create these discoverers is often manual but can also be semi-automatic. Result of the MoDisco process is an UML file in XML format.

- Understanding the extracted information in order to take the good modernization decisions: The UML model produced in MoDisco is a large data file where not all data are meaningful and useful. In this phase parsing of the XML file is needed to extract only data like objects, attributes and relations that would be useful information that sufficiently describe artefacts the of existing system. The extraction process requires a custom code to be written. Result of this phase is an intermediary XML file, which is a restructured subset of the original UML file. This phase is iterated over all existing university information systems (in our case three—UM, UCC, DIT).

- Transforming this information to new artefacts facilitating the modernization: If previous phase resulted in many different XML files, then these files needs to evolve into a single model acceptable to all initial participating universities. The task is a semi-manual activity, which also includes negotiation meetings. This is the moment where the basic taxonomy for inter-university collaboration from Sect. 4.2.2 needs to be taken into account. In this phase similarities of the syntax and semantics of the existing meta-models are programmatically analyzed and compared. Final result is a model compromise suitable for HLA FOM file generation. Another result of this phase is also ontology in format Ontology Web Language (OWL) as shown on the Fig. 4.9.

- RTI federated platform setup: The HLA FOM file from the previous phase is actually generated with the HLA platform requirements in mind. Our choice for HLA platform is the open source software poRTIco [51] which supports the RTI FOM format. Portico is a cross-platform (supports Java, C++) HLA RTI implementation. Designed with modularity and flexibility in mind, Portico is intended to provide a production grade RTI implementation and an environment that can support continued research and development. The goal of this phase is to get a federation up and running. As shown on Fig. 4.8 the three UISs and the ITCEM VLE have been converted to RTI Federates thru the modernization process. The Federates are code plug-ins to the poRTIco provided platform and serve as syntax and semantics (behavior) mediators between existing and unchanged IS.

- New ITCEM members, outside of the federation, can join the federation thru the WebService Federate. New ITCEM members are typically universities that want to join the ITC Euromaster programme as content providers and/or consumers. These new members are not forced to go through the model alignment process

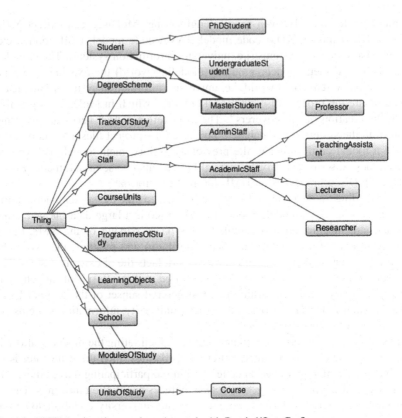

Fig. 4.9 ITCEM university ontology (created with Protégé/OntoGraf)

because that would ruin the execution of the existing federation. Therefore the WebService Federate must tackle syntax and semantics disharmony between the federation and the new member. To address that disharmony we propose the use of intelligent agents and ontology. Details are described in the next Sect. 4.4.1.

4.4.1 Resolving Disharmony Between Federation and New ITCEM Members

In order to plug-in the new ITCEM member via WebService Federate to the existing ITCEM Federation, semantical and syntactical disharmony issues needs to be resolved first (authors have done similar research in the past, however applied in the field of construction [52, 53]). While the ITCEM Federation already possess its own ontology as shown on Fig. 4.9, the new ITCEM member needs to build up their own local ontology for presentation Fig. 4.9 shows the ITCEM University ontology developed according to the taxonomy for inter-university collaboration

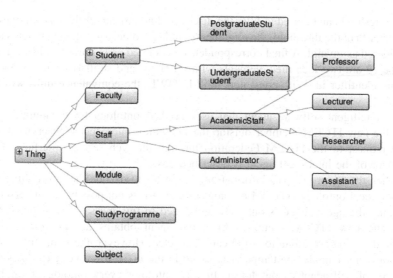

Fig. 4.10 New ITCEM member ontology (created with Protégé/OntoGraf)

on Fig. 4.2. Figure 4.10 shows the ontology of a new ITCEM member university that connects to ITCEM Federation thru the WebService Federate. Figures 4.9 and 4.10 illustrate the problem of ontology alignment.

The multiple ontology approach is supportive for achieving federated approach because multiple ontology approach has no ontology commitment about shared ontology. Each information source is described by its own local ontology. Federated approach requires dynamical adjustment and alignment without predefined common format.

In order to find equivalent concepts between the ontologies of existing HLA federation and new participant, ontology matching will be performed with a multi-strategies-based approach (as proposed by [54]). In this approach, two source ontologies are the inputs.

Algorithm description for ontology alignment process is as follows [54]:

- A pre-process will be carried out to eliminate and tokenize source ontology into single elements.
- For each pair of elements, a strategy will be applied to select one or more suitable matchers. There are three matchers used in the approach from different aspects of source ontology: string, structural and semantic. Each selected matcher will generate one similarity value.
- In order to aggregate different matching results, an analytic method with Analytic Hierarchy Process (AHP) is adopted to learn the weight of each matcher. The process is based on three similarity indicators, which could reflect the essential features of source ontology, to assign the intensity of importance when measuring the criteria against the goal. A final correspondence will be generated with the learned weights.

A threshold can be used to filter the discovered alignments. When the similarity is greater than the threshold, the alignments are kept, otherwise, the alignments are considered as invalid. A final correspondence is defined as {e1, e2, r, v, id}, where e1 and e2 are two identified elements with relation r and similarity value v and a unique identifier id. With constructs built-in OWL, the equivalence links will be setup.

An intelligent software agent that uses ITCEM ontology for communication with the new ITCEM member outside the federation intercepts Webservice Federate request to the ITCEM Federation through the web service interface. The behavior of the intelligent agent is described next.

When the agent receives information, it will try to decode the information by using a local ontology (ITCEM ontology). In case it is not possible to understand the data, the agent requests ontology associated to this message to be delivered from the new ITCEM member. After the agent obtains all the information required, the received ontology file can be deleted. However, the terms inside the ontology files can also be temporarily saved in the ITCEM ontology (follows the concept of self-adaptive ontology). In this ontology every ontology term has weighting coefficient, which can measure the popularity of ontology concept. In case the ontology concept gets a low coefficient, it will be deleted from ontology.

According to the requirements of the federated architecture, this approach supports "on-the-fly" (plug-and-play) scenario within the ITCEM Federation.

4.5 Requirements for Further Development of the ITCEM

The ITCEM Curse Pool in this form has been in use since 2006. In 2012 the new ITCEM site (http://euromaster.itcedu.net) has been launched. In 2012 new institution (University of Dresden) expressed its interest to provide a new course to the pool. Authors' experiences gained with the management and coordination of the ITCEM in the past years identified following requirements for improvement:

- Need for common vocabulary to resolve ambiguity: In the United Kingdom (i.e. Ireland) "course" refers to the entire programme of studies required to complete a university degree, and the word "unit" or "module" would be used to refer to an academic course in the North American sense [55]. In Germany and Slovenia "course" refers to the unit of study equal to the "module" in Ireland. In Slovenia "module" is a specialized track while in Ireland it refers to unit of study. Therefore the taxonomy TOSU for ITCEM needs to be further developed and computerized to define synonyms. Common vocabulary offers advantages for teachers and learners. The taxonomy support within the LMS should be examined. Modern LMS portals have plugged-in taxonomy modules that enable flat (as metadata in tagging system) or hierarchical arrangement (with parents and children) implementation of taxonomies.

- Further development of semantic linking between ITCEM and external resources: Semantic web technologies (Web 3.0) and ontology mapping/ alignment this would provide a higher level of decision support analysis and mining, based in qualitative issues like: the pedagogic methodologies used, the collaborative degree of activities or the understanding expressed in the assessments and assignments.
- Automatic synchronization of timetables between ITCEM LMS and corresponding university ISs: Bi-directional propagation of timetable changes.
- Automatic enrollment with ITCEM LMS: Manual student enrollment process to ITCEM courses should be upgraded to transfer student information from existing university-wide student management system. This would avoid duplicate enrollment, which is currently in place.
- Automatic synchronization of assessment results between ITCEM LMS and corresponding university ISs: On-request secured delivery of exam results from LMS to the requesting university for the needs of student records services.
- Implementation of adaptable e-learning services within the LMS: In order to achieve this, we would like to collect data on student motivation with the online activities, student satisfaction with the online learning environment, student online activity statistics from LMS (Moodle) and final course grades [56]. Similar approaches like personalized e-learning systems based on intelligent agents [32], e-learning personalization [29] and adaptable e-learning services [34, 57] are active research topics.
- Adherence to emergent e-learning standards: Many organizations like IMS Global Learning Consortium (http://www.imsglobal.org), IEEE LTSC (http:// ltsc.ieee.org), ADL (http://www.adlnet.org), ARIADNE and AICC are making standards in the field of e-learning and most of the standards are becoming the de-facto standards in e-learning [58]. These standards have been defined to structure learning by also providing metadata to represent its objects (e.g. multimedia content, instructional content, learning objectives, instructional software, learner profiles, etc.).

We hope to achieve all the other of the above requirements with further study, development and adaptation of the existing virtual learning environment towards systems such as adaptive intelligent educational systems and adaptive web based e-learning systems [30]. We believe that we can achieve this with a semantic web technologies-based multi-agent system that allows to automatically control students' acquired knowledge in e-learning environment. Such systems allowing the integration not only of intelligent agents that support the teaching/learning process, but also adapting the learning process to each particular student. In order to do that, such systems try to analyze the students' interactions and, if possible, their solutions to assignments. The analysis of students' mistakes allows to propose them personalized recommendations and to improve the course materials in general. In that way, they try to determine the student cognitive stage so that the learning process can be adapted to each concrete student.

4.6 Conclusions

The main objective of the ITCEM was (and still is) to organize the knowledge in the field of IT in AEC and to develop an effective environment to support "collaborative learning scenarios" with distributed students and teachers. The establishment of a community spirit bears at least the same importance for the success of the developed programme as the implementation, configuration and further development of the virtual learning environment. The ITCEM VLE with its High Level Architecture will become the basis for a virtual university, linking together teachers and students of multiple nations and continents and independent of the heterogeneous technical facilities they connect from to the ITCEM Federation. For this reason, we will continue to develop new content for the ICT-Euromaster course pool and strive to improve all presented state-of-the-art technical aspects.

We are convinced that the civil and building industry will need more engineers with profound IT understanding and knowledge in the e-society of tomorrow.

References

1. Rebolj, D., Menzel, K., Dinevski, D.: A virtual classroom for information technology in construction. Comput. Appl. Eng. Educ. **16**, 105–114 (2008). doi:10.1002/cae.20129
2. Tibaut, A., Rebolj, D., Čuš Babić, N.: The past and the future of the European master's programme in information technology in construction. In: Electronic proceedings 14th International Conference on Computing in Civil and Building Engineering (14th ICCCBE), Moscow State University of Civil Engineering, Moscow, Russia, June 27-29, V (2012)
3. Tibaut, A., Rebolj, D.: ITC Euromaster. http://euromaster.itcedu.net. Accessed 15 Jan 2013
4. Tibaut, A., Menzel, K.: International e-learning awards, honorable mention, e-learning. http://www.ielassoc.org/awards_program/past_winners.html (2012). Accessed 15 Jan 2013
5. Easterby-Smith, M., Lyles, M.A., Tsang, E.W.K.: Inter-organizational knowledge transfer: current themes and future prospects. J. Manage. Stud. **45**, 677–690 (2008). doi:10.1111/j.1467-6486.2008.00773.x
6. Semradova, I.: Designing e-learning courses in humanities and their use in the interuniversity study programmes. Procedia Comput. Sci. **3**, 162–166 (2011). doi:10.1016/j.procs.2010.12.028
7. Hubackova, S.: Possibilities of the use of ICT in interuniversity studies. Procedia Soc. Behav. Sci. **28**, 29–33 (2011). doi:10.1016/j.sbspro.2011.11.006
8. Van der Wende, M. C.: Internationalization of Higher Education. In: Peterson, P., Baker, E., McGaw, B. (eds.) International Encyclopedia of Education, 4, pp. 540–545. Elsevier, Oxford, UK (2010)
9. Knight, J.: Higher Education Crossing Borders. In: Baker, E., Peterson, P., McGaw, B. (eds.) International Encyclopedia of Education 3rd Edition, pp. 507–513. UK Elsevier Publishers, Oxford (2010)
10. Sutrisno, A., Hartanto, D., Pillay, H.: Transnational higher education programs for facilitating inter-university knowledge transfer: university of Indonesia's experience. In: ISANA International Education, pp. 1–10 (2012)
11. University of Minnesota, Inter-College Programme. http://www.cce.umn.edu/Inter-College-Program/. Accessed 15 Jan 2013

12. Dado, E., Beheshti, R.: Digital learning environments for joint master in science programmes in building and construction in Europe: experimenting with tools and technologies. In: Proceedings of World Academy of Science: Engineering and Technology, pp. 440–447 (2009)
13. Uchiyama, K.P., Radin, J.L.: Curriculum mapping in higher education: a vehicle for collaboration. Innovative High. Educ. **10**, 10–20 (2008)
14. Bruno, J.E.: The use of Monte Carlo techniques for determining optimal size of substitute teacher pools in large urban school districts. Socio-Economic Plann. Sci. **4**(4), 415–428 (1970). doi:10.1016/0038-0121(70)90018-2
15. Stricker, D., Weibel, D., Wissmath, B.: Efficient learning using a virtual learning environment in a university class. Comput. Educ. **56**, 495–504 (2011). doi:10.1016/j.compedu.2010.09.012
16. Helokunnas, T., Herrala, J.: Knowledge searching and sharing on virtual networks. In: Proceedings of the ASIST Annual Meeting, vol. 38, pp. 315–322 (2001)
17. Rosen, B., Furst, S., Blackburn, R.: Overcoming barriers to knowledge sharing in virtual teams. Organ. Dyn. **36**, 259–273 (2007). doi:10.1016/j.orgdyn.2007.04.007
18. Alsharo, M.K.: Knowledge sharing in virtual teams: the impact on trust, collaboration, and team effectiveness, PhD diss., University of Colorado (2013)
19. He, J.: A Examining factors that affect knowledge sharing and students' attitude toward their learning experience within virtual teams. Dissertation Abstracts International A, The Humanities and Social Sciences, 71 (2009)
20. Kleanthous Loizou, S.: Intelligent Support for Knowledge Sharing in Virtual Communities. University of Leeds, UK (2010)
21. Usoro, A., Sharratt, M.W., Tsui, E., Shekhar, S.: Trust as an antecedent to knowledge sharing in virtual communities of practice. Knowl. Manage. Res. Pract. **5**, 199–212 (2007). doi:10.1057/palgrave.kmrp.8500143
22. Majewski, G., Usoro, A., Khan, I.: Knowledge sharing in immersive virtual communities of practice. Vine **41**, 41–62 (2011). doi:10.1108/03055721111115548
23. Bertino, E., Squicciarini, A.: A decentralized approach for controlled sharing of resources in virtual communities. 10th international conference on computer supported cooperative work in design, pp. 1–1 (2006). doi:10.1109/CSCWD.2006.253000
24. Li, W.: Virtual knowledge sharing in a cross-cultural context. J. Knowl. Manage. **14**, 38–50 (2010). doi:10.1108/13673271011015552
25. Ray, F.D.P., Rabhi, F.: Knowledge Sharing Infrastructure for Virtual Enterprises. In: Khosrowpour, M. (ed.) Information Technology and Organizations Trends Issues Challenges and Solutions, vol. 1, p. 307. University of South Wales, Australia (2003)
26. De Meo, P., Garro, A., Terracina, G., Ursino, D.: Personalizing learning programs with X-learn, an XML-based, "user-device" adaptive multi-agent system. Inf. Sci. **177**, 1729–1770 (2007). doi:10.1016/j.ins.2006.10.005
27. Yang, F., Wang, M., Shen, R., Han, P.: Community-organizing agent: an artificial intelligent system for building learning communities among large numbers of learners. Comput. Educ. **49**, 131–147 (2007). doi:10.1016/j.compedu.2005.04.019
28. Tien, L.T., Osman, K.: Pedagogical agents in interactive multimedia modules: issues of variability. Procedia Soc. Behav. Sci. **7**, 605–612 (2010). doi:10.1016/j.sbspro.2010.10.082
29. Klašnja-Milićević, A., Vesin, B., Ivanović, M., Budimac, Z.: E-Learning personalization based on hybrid recommendation strategy and learning style identification. Comput. Educ. **56**, 885–899 (2011). doi:10.1016/j.compedu.2010.11.001
30. Jurado, F., Redondo, M., Ortega, M.: Blackboard architecture to integrate components and agents in heterogeneous distributed e-learning systems: an application for learning to program. J. Syst. Softw. **85**, 1621–1636 (2012). doi:10.1016/j.jss.2012.02.009
31. Latham, A., Crockett, K., McLean, D., Edmonds, B.: A conversational intelligent tutoring system to automatically predict learning styles. Comput. Educ. **59**, 95–109 (2012). doi:10.1016/j.compedu.2011.11.001

32. Duo, S., Ying, Z.C.: Personalized e-learning system based on intelligent agent. Phys. Procedia **24**, 1899–1902 (2012). doi:10.1016/j.phpro.2012.02.279
33. Huang, W., Webster, D., Wood, D., Ishaya, T.: An intelligent semantic e-learning framework using context-aware semantic web technologies. Br. J. Educ. Technol. **37**, 351–373 (2006). doi:10.1111/j.1467-8535.2006.00610.x
34. Sarraipa, J., Baldiris, S., Fabregat, R., Jardim-Goncalves, R.: Knowledge representation in support of adaptable e-learning services for all. Procedia Comput. Sci. **14**, 391–402 (2012). doi:10.1016/j.procs.2012.10.045
35. Sejzi, A.A., Aris, B., Yahya, N.: The phenomenon of virtual university in new age: trends and changes. Procedia Soc. Behav. Sci. **56**, 565–572 (2012). doi:10.1016/j.sbspro.2012.09.689
36. Shahtalebi, S., Shatalebi, B., Shatalebi, F.: A strategic model of virtual university. Procedia Soc. Behav. Sci. **28**, 909–913 (2011). doi:10.1016/j.sbspro.2011.11.167
37. Yengin, İ., Karahoca, D., Karahoca, A., Uzunboylu, H.: Re-thinking virtual universities. Procedia Soc. Behav. Sci. **2**, 5769–5774 (2010). doi:10.1016/j.sbspro.2010.03.941
38. Zheng, A.Y., Lawhorn, J.K., Lumley, T., Freeman, S.: Assessment. Application of Bloom's taxonomy debunks the "MCAT myth". Science **319**, 414–415 (2008)
39. Jansen, B.J., Booth, D., Smith, B.: Using the taxonomy of cognitive learning to model online searching. Inf. Process. Manage. **45**, 643–663 (2009). doi:10.1016/j.ipm.2009.05.004
40. Dalkir, K.: Knowledge Management in Theory and Practice. Butterworth-Heinemann, Burlington, MA (2005)
41. Jardim-Goncalves, R., Grilo, A., Agostinho, C., et al.: Systematisation of interoperability body of knowledge: the foundation for EI as a science. Enterp. Inf. Syst. (2012). doi:10.1080/17517575.2012.684401
42. Charalabidis, Y., Gionis, G., Hermann, KM., Martinez, C.: Revision of the enterprise interoperability research roadmap (v 5.0) (2011). ftp://cordis.europa.eu/pub/fp7/ict/docs/enet/ei-roadmap-5-0-draft_en.pdf
43. Bourey, J-P., Reyes, G., Doumeingts, G., Berre, JA.: DTG2.3. Report on model driven interoperability—I–V Lab Platform (2007)
44. ADM/KDM version 1.2 (2010). http://www.omg.org/spec/KDM/1.2/
45. ASTM version 1.0 (2011). http://www.omg.org/spec/ASTM/
46. IEEE Standard for Modeling and Simulation, High Level Architecture (HLA) (2010). doi:10.1109/IEEESTD.2010.5557731
47. Moore, J.L., Dickson-Deane, C., Galyen, K.: E-Learning, online learning, and distance learning environments: are they the same? Internet High. Educ. **14**, 129–135 (2011). doi:10.1016/j.iheduc.2010.10.001
48. Wood, W.B.: Sharing in the classroom. CBE Life Sci. Educ. **7**, 263–264 (2008)
49. Tu, Z., Zacharewicz, G., Chen, D.: Developing a web-enabled HLA federate based on portico RTI. In: Proceedings of the Winter Simulation Conference, pp. 2294–2306 (2011)
50. MoDisco v.0.10.0. Eclipse.org, http://www.eclipse.org/MoDisco/
51. The Portico Project (2013). Sourceforge.net, http://www.porticoproject.org/
52. Tibaut, A.: Intelligent agents for better information management process in construction. In: Proceedings of CIT2000—The CIB-W78, IABSE, EG-SEA-AI International Conference on Construction Information Technology (2000)
53. Tibaut, A.: Resolving disharmony in structurally and semantically heterogeneous information systems, PhD thesis, University of Maribor, p. 146 (2002)
54. Song, F., Zacharewicz, G., Chen, D.: Multi-strategies Ontology Alignment Aggregated by AHP. In: Frontiers in Artificial Intelligence and Applications. Advances in Knowledge-Based and Intelligent Information and Engineering Systems, 243, pp. 1583–1592. (2012). doi: 10.3233/978-1-61499-105-2-1583
55. Wikipedia. Wikipedia: Course (education). http://en.wikipedia.org/wiki/Course_(education) (2013). Accessed 15 Jan 2013

56. Cuadrado-García, M., Ruiz-Molina, M.-E., Montoro-Pons, J.D.: Are there gender differences in e-learning use and assessment? Evidence from an interuniversity online project in Europe. Procedia Soc. Behav. Sci. **2**, 367–371 (2010). doi:10.1016/j.sbspro.2010.03.027
57. Vesin, B., Ivanović, M., Klašnja-Milićević, A., Budimac, Z.: Ontology-Based Architecture with Recommendation Strategy in Java Tutoring System. Comp. Sci. Inf. Sys. 10, pp. 237–261 (2013)
58. Akshay, B., Mumbai, C.: Specifications and standards in e-learning. http://www.cdacmumbai.in/design/corporate_site/override/pdf-doc/Specifications_and_Standards_in_E-Learning.pdf (2009). Accessed 15 Jan 2013

Chapter 5
An Agent Based E-Learning Framework for Grid Environment

Sarbani Roy, Ajanta De Sarkar and Nandini Mukherjee

Abstract This chapter presents how Grid can be used to build an e-learning framework which is flexible, convenient, cost-effective, and adaptable. Grid technologies are appealing due to the fact that the requirements for developing such framework match very closely what a Grid can offer in terms of computational and storage resources. On the other hand, agent-based technology can make e-learning Grid more efficient. A few autonomous, co-operative agents with predefined functionalities and responsibilities provide more powerful and reliable e-learning system. The objective of this chapter is to accomplish a blended e-learning Grid framework, where the framework is designed as a multi-agent system integrated with Grid. This chapter also presents the implementation of an e-learning system as Grid services and analysis of the benefits of e-learning Grid system.

Keywords E-learning · Grid · Multi-agent

5.1 Introduction

In recent years, there has been a growing interest to reduce costs of establishing learning environment systems. Modern e-learning systems are envisioned as adaptable, interactive, distributed and collaborative systems. Over the Internet,

S. Roy (✉) · N. Mukherjee
Department of Computer Science and Engineering, Jadavpur University,
Kolkata 700032, India
e-mail: sarbani.roy@cse.jdvu.ac.in; sarbani.roy@gmail.com

N. Mukherjee
e-mail: nmukherjee@cse.jdvu.ac.in

A. De Sarkar
Department of Computer Science and Engineering, Birla Institute of Technology, Mesra,
Kolkata Campus, Kolkata 700107, India
e-mail: adsarkar@bitmesra.ac.in

M. Ivanović and L. C. Jain (eds.), *E-Learning Paradigms and Applications*,
Studies in Computational Intelligence 528, DOI: 10.1007/978-3-642-41965-2_5,
© Springer-Verlag Berlin Heidelberg 2014

learners can effectively realize the learning process at any time from anywhere. Generally, e-learning systems are designed using client/server, peer to peer; and recently web service architectures. These systems have major drawbacks because of their limitations in scalability, availability, distribution of computing power and storage system, as well as sharing information between users that contribute in these systems. Therefore, there is a need to redesign the e-learning system to meet the needs better. Moreover, complex applications which are computationally intensive and handle large data sets have been ignored in the context of e-learning up to now, mainly due to technical feasibility problems and prohibitively high costs. Grid computing can close this gap and enable new types of e-learning applications. Computations and data can be distributed on Grid if local computers fail to handle them. Using the concept of virtual organization in Grid, users and organizations can be effectively grouped for cooperative learning. Hence, e-learning systems can be extended to e-learning Grids in which Grid computing functionalities are integrated.

On the other hand, through an agent-based approach, the ideas of personalization and interactive learning can also be incorporated more easily. The opportunities for using agents in e-learning applications are enormous. Introducing agents into the e-learning environment will fundamentally change the way online education is conducted. Agent characteristics like autonomy, abilities to perceive, reasoning and ability to act in specialized domains, as well as their capability to cooperate with other agents makes them ideal for e-learning applications. Each agent contains its own functionalities and responsibilities to achieve its own objectives. Also, an agent can assist other agents when they are in need of help in order to complete the task or request. Agents can also monitor learning effectiveness, and thus the benefits of the e-learning program can be assessed by the organization. Moreover, the mobile agent technology is particularly suitable for developing distributed e-learning systems because it enhances modularity, reusability, flexibility and reliability. A mobile agent helps to change remote interaction into local interaction. It has some advantages such as less dependent on network, allowing network interruption, reducing network occupying time, overcoming network delay, increasing the utilization ratio of network, and improving the response speed of users interaction request. Users can search for suitable learning objects with agent assistance using Open Grid Service Infrastructure (OGSI) compliant interfaces. Data sets of learning objects are managed and controlled by agents.

There is a huge potential of Grid and multi-agent systems to enhance each other because these models have developed significant complementarities. Multi-agent systems need a robust distributed computing environment that helps them to discover, acquire, federate, and manage the capabilities necessary to execute their decisions [1]. E-learning Grid provides a powerful framework for matching needs and capabilities and for combining capabilities to address those needs by leveraging other capabilities. The Grid based e-learning framework improves the learning process; simultaneously exploit huge computational power and data storage. The modern e-learning system indeed needs huge data, like presentation slides, images, animations, videos and many other media files. So, the learning resource management and the maintainability and expandability of an e-learning system are very essential. In this highly dynamic and heterogeneous environment, autonomy and flexibility is essential. Hence, multi-agent based e-learning Grid can move forward a robust distributed computing environment to discover, acquire, federate, and manage the capabilities necessary to execute their services.

This chapter will focus on a framework for agent based e-learning Grid. The architecture of e-learning Grid encapsulates educational materials inside Grid services which satisfy the demands of interoperability and reusability. On the other hand, Grid core services are used for resource sharing. Agents are created, deployed and published as web services. Agents are responsible for management of distributed learning resources, discovery and selection of learning services, setting up service level agreement (SLA), service level monitoring etc. Depending on the role of these agents, some are acted as mobile agents. To accomplish the above mentioned tasks a middleware is required that will integrate Grid and agent technologies. The middleware for e-learning Grid is designed as a multi-agent system and built as a set of different middleware layers, which work together to provide transparent resource sharing environment for the said application.

The remainder of this chapter is organized as follows. Section 5.2 reviews related work. An overview of Grid environment is presented in Sect. 5.3. The concepts related to Grid middleware is also discussed in this section. Different components of e-learning system and their working procedures are discussed in Sect. 5.4. Architecture of e-learning Grid system is presented in Sect. 5.5. Design of e-learning Grid as multi-agent system is presented in Sect. 5.6. Implementation of an e-learning service as a Grid service is discussed in Sect. 5.7. Section 5.8 analyses the benefits of e-learning Grid system. Concluding remarks are given in Sect. 5.9.

5.2 Related Work

This section presents review of a few existing research on e-learning or e-learning Grid system. The architectures that surveyed here gives users the ability to collect, analyze, distribute and use e-learning knowledge from multiple knowledge sources.

In [2], web service-based framework for e-learning portal system is proposed. This system provides an environment to present collaborated e-learning by facilitating efficient communication in components and portal. Web service based thin client architecture for e-learning system is proposed in [3] that uses Run Time Environment (RTE) in SCORM to trace learning process with a suitable middleware component. Su et al. [4] discussed a set of e-learning web services include assessment, course management, grading, marking, Metadata, registration and reporting web services. The proposed architecture is based on e-learning framework (ELF) and consists of four main layers: presentation, e-learning services, common services and resources. ELF is a service-oriented featuring of the core services required to support e-learning applications, portals.

Nowadays, modern e-learning system has met challenges in learning resources or services sharing and reuse, interoperability. These challenges can be tackled by Grid technologies. European learning grid infrastructure (ELeGI) project aims to address and advance present e-learning solutions through collaborative use of geographically distributed computing and educational resources as a single e-learning environment [5]. In [6], Grid-enabled large-scale collaboration environment (GLCE) is proposed. Grid-based cooperative work framework (GCWF) is also proposed to realize GLCE and used to build the learning assessment Grid (LAGrid). In [7], distributed architecture for dynamic e-learning environment (RDADeLE) is proposed. This architecture incorporates web services, and agent technology. Moreover, it uses regional Data Grid for accessing vast learning materials of e-learning system. This integrated multi-agent based dynamic e-learning system is expected to be scalable, stronger and efficient architecture. A layered architecture to manage data, information and knowledge to enhance the education system is presented in [8]. In [9], an e-learning middleware is proposed including agents, web services and Grid technology. Except the basic e-learning services, this middleware actually uses core Grid services as additional e-learning services. In [10], the composite state goal model is developed as agent mediated e-learning in Grid environment with Marketing Agents and Service Agents hierarchically. These two agents can negotiate each other to represent service providers and service consumers. Shen et al. developed this agent model in different hierarchical level with different sets of agent to reach common goal. However, [10] does not specifically explain any agents involved with core Grid services.

Unlike other systems, the proposed system not only be used as the traditional e-learning system, but it also provides benefit to the remote users dynamically through virtual organizations. Moreover, this proposed system uses data Grid as a huge storage for fulfilling the demand of the users easily. Generally, individuals and organizations possess limited resources. Grid based e-learning framework can provide support for dynamism in terms of resources, content and participants. The proposed system considered them as a core Grid service feature to design effective E-learning system.

5.3 Grid as Infrastructure

During the last two decades, an evolution in the Internet technology has occurred. Alongside, powerful computers and high-speed network technologies have become available as low-cost commodity components and combination of these two has changed the way we use computers today. These technology opportunities have led to the possibility of using distributed computers as a single, unified computing resource, leading to what is popularly known as Grid computing. The Grid computing environment is very appropriate for the e-learning application since it opens new potentials of improving the learning process. The advantages that Grid architecture offers flexible and coordinated way of sharing resources in the Internet as well as on its enormous capabilities of information processing increased their demand.

5.3.1 Grid Computing

The idea of Grid has been conceived as a large scale, generalized distributed network computing system that can scale to Internet-size environments with machines distributed across multiple organizations and administrative domains. The concept of Grid has evolved from the idea of fulfilling the resource requirement of an application with the help of coordinated resource sharing among dynamic collections of individuals, institutions, and resources. This sharing is, necessarily, highly controlled, with resource providers and consumers defining clearly and carefully what is shared, who is allowed to share, and the condition under which sharing occurs. A set of individuals and/or institutions defined by such sharing rules form a group called virtual organization (VO) [11]. Thus, creating virtual organizations and enterprises as a temporary alliance of enterprises or organizations that come together to share resources and skills, core competencies, or resources can be enhanced (multiplied) in order to better respond to business opportunities or to fulfill large-scale application processing requirements as envisioned in [12]. The cooperation among the VOs is supported by computer networks.

The mid-level software that provides services to users and to the applications in the Grid environment is called middleware. Next section highlights the importance of Grid middleware and describes the architecture and the infrastructure to be supported by any particular Grid middleware.

5.3.2 Grid Middleware

Grid middleware are software stacks designed to present different compute and data resources in a uniform manner. The middleware stack is a series of cooperating programs, protocols and agents designed to help users accessing the resources of the Grid. An appropriate Grid middleware must provide a reliable implementation of the fundamental Grid services, such as information services, job management services which include job submission and monitoring, resource management services which include resource discovery, monitoring and brokering and data management services. The complexity of accessing the hardware and the software resources are entirely managed by the Grid middleware. Thus, the intention of the Grid middleware is to create virtual organizations with the resources, provide access while maintaining the policies at different levels, and in general deal with the physical characteristics of the Grid. In order to reduce complexity, Grid middleware should allow users and applications to access Grid resources in a transparent manner, the user does not need to know where the resource is physically located, and the type of machine it is on. From a user's perspective, therefore, Grids are all about resource accessing across sites and organizations without dealing with complexities related to management and security policies, accessibility options and other such type of issues.

The existing Grid middlewares are built as layered interacting packages and controlled by different managers called by a common API. The users are not concerned with the different syntaxes and access methods of specific packages. The user can simply submit its job through the API to the job manager.

The Globus Toolkit [13], developed by the Globus Alliance, is a Grid middleware constructed from a number of components that make up a toolkit. Globus includes quite a few high-level services that can be used to build Grid applications. The Open Grid Services Architecture (OGSA) [14] developed by the Global Grid Forum (GGF) [15], aims at defining a common, standard, and open architecture for Grid-based applications. The objective of OGSA is to standardize practically all the services, which are commonly used for executing an application in Grid environment [16]. These include job management services, resource management services, security services, etc. OGSA specifies a set of standard interfaces and depends on web services as the underlying middleware. OGSA first spawned the Open Grid Services Infrastructure (OGSI) [17], which, despite the improvements of web services in several ways, failed to converge with existing web services standards. Although the web services architecture was certainly the best option, but it still failed to meet one of the most important requirements of OGSA—the underlying middleware had to be stateful. Web services, though in theory can be both stateless or stateful, they are usually stateless and there is no standard way of making them stateful. With an objective to fulfill this need, a new set of specifications for web services (published by OASIS [18]) have been introduced with major contributions from the Globus Alliance and IBM. These specifications are defined as Web Services Resource Framework (WSRF) [19] and form the infrastructure on which OGSA the Grid architecture is built on. WSRF defines web service conventions to enable the discovery of stateful services [20] and interactions with them in standard and interoperable ways. Thus, WSRF provides the stateful services that OGSA needs. WSRF basically improved on OGSI and eventually replaced it. Globus toolkit includes quite a few high-level services that can be used to build Grid applications. These services, meet most of the abstract requirements set forth in OGSA. Most of these services are implemented on top of WSRF.

The Globus toolkit provides client, server and development components for the three Globus "Pyramids" [14] of Grid Computing: Resource management, Information Management and Data Management. Globus Toolkit has these modules built on top of a security infrastructure, namely Grid Security Infrastructure (GSI). These three modules provide support for resource management, data management and information management and GSI provides different security services like authentication and authorization. Globus is implemented with computers on a Grid that are networked and running applications. They also handle sensitive or extremely valuable data, therefore the security component of Grid plays an important role. In addition, Globus implements the resource management, data management and information services as separate components. The resource management component provides support for resource allocation, job submission, managing job status and monitoring progress of the job. Information services provide support for collecting static and dynamic information about resources and for querying this information. This is based on the Lightweight Directory Access Protocol (LDAP). The data management component provides support for transferring files among machines in the Grid and for managing these transfers.

5.3.3 Service Oriented Architecture

Service Oriented Architecture (SOA) is very much useful to build and enhance Grid middleware services and is widely seen as a base for new models of distributed applications (e.g., e-learning). SOA [21, 22] is a prototype for designing, organizing and utilizing distributed functionalities that may be under the control of different ownership domains and implemented using various components or services that are used to provide a software solution or application.

One of the main aspects of service-oriented architecture is that it separates the services implementation from its interface. In SOA, a service is viewed by service consumer as an endpoint that supports a particular request format or contract. The way the service executes tasks given to it is irrelevant. The most basic message exchange pattern is a common Request-Response where the parties can simply communicate with each other. Request-Response is a pattern in which the service consumer uses configured client software to issue an invocation request to a service provided by the service provider. A service registry can also be used within the architecture to help the consumer automatically configure certain aspects of its service client. The service provider publishes its contract in the registry for access by service consumers. A service registry is an entity that accepts and stores contracts from service providers and provides those contracts to interested service consumers. A contract is a specification of the way a consumer of a service will interact with the provider of the service. It specifies the format of the request and response from the service. The contract may also specify quality of service (QoS) levels. This is conceptually represented in Fig. 5.1.

The definition of SOA is technology independent. However the most widely used implementation of SOA is in web Services, where services are communicating using established Internet standards. The language for describing web services is based on XML and is called Web Services Description Language (WSDL). In general, Simple Object Access Protocol (SOAP) is the standard messaging protocol used by web services. SOAPs primary application is Application-to-Application (A2A) communication. Universal Description, Discovery, and Integration (UDDI) provides a registry mechanism for service consumers and service providers to find each other and uses SOAP for communication.

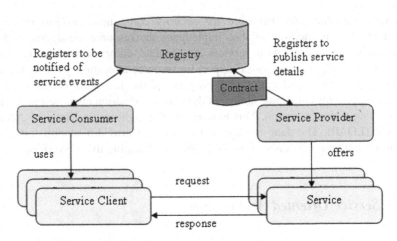

Fig. 5.1 SOA data and message exchange pattern with a service registry

5.4 E-Learning System

By nature e-learning system is a web application. The quantity of data in modern e-learning system is huge. Presentation slides, images, animations, videos and many other media files are now becoming important learning resources. So, the learning resource management and the maintainability and expandability of an e-learning system are very essential. To develop e-learning systems, many researchers made their research on the architecture and component of e-learning systems, which include workflow based architecture, knowledge flow driven architecture, P2P based architecture, service oriented architecture, web-service based architecture etc [3, 23, 24, 25]. A multi-layered framework for designing an e-learning system is illustrated in Fig. 5.2. A generic view of e-learning system architecture is depicted in Fig. 5.3. The architecture of e-learning system must be able to integrate the services of each layer. Following is the brief description of each layer.

Fig. 5.2 Multi-layered framework of e-learning system

Layer 1: E-learning Portal User access
Layer 2: User Management Common services
Layer 3: Learning Management Learning services/Content development & delivery
Layer 4: Data Storage & Access XML/PostgreSQL/File access/File sharing
Layer 5: Infrastructure Internet/HTTP/SOAP/FTP/SMTP/TCP-IP

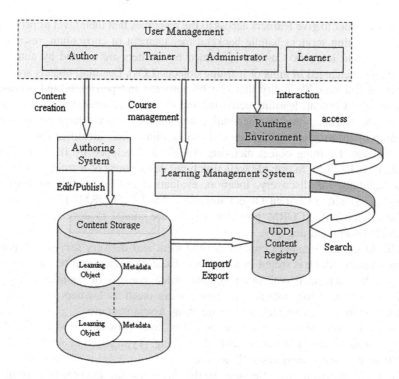

Fig. 5.3 Generic view of e-learning system architecture

Layer 1: *E-learning Portal*—It is a single entry point that allows all users to access all relevant part of the system via a standard web browser. The log-in takes the user into the user management service for authentication.

Layer 2: *User Management*—This layer mainly provides services that are needed by every user and are not tied to any particular pedagogic function. Each user is identified with a unique ID, to which roles can be assigned with distinct privileges. Roles include: author, trainer, administrator and learner, etc. Roles and privileges can be changed as often as desired. Individuals can have multiple roles and therefore numerous privileges may be assigned to them. The user management service records and handles all the user information and conducts the authentication process. This layer also provides collaboration services to establish communication among all users of the system. Both synchronous and asynchronous collaboration technologies are implemented. Virtual classroom (with audio-visual, and whiteboard resources), virtual meeting rooms, and chat are examples of synchronous collaboration. Asynchronous collaboration technologies include email, threaded discussion, and peer-to-peer instant messaging.

Layer 3: *Learning Management*—This layer provides services to manage core functionality for the production and consumption of e-learning resources. The learning management system (LMS) is tightly connected to the user management services in order to give learners access to the resources that their level of privilege allows. Learning services handle backend management of curriculum, resources, instructors and learners. Content consumed by learners are created by author is stored and exchanged in units of learning objects (LOs). Learning objects are self-contained and reusable entity that can be authored independently and accessed dynamically. Content, learning activities and elements of context are three components of learning objects. Generally, the learning objects have an external structure of information called metadata to facilitate their identification, storage and retrieval. Learning object metadata (LOM) is developed by IEEE Working Group P1484.12 [26] to provide well-structured descriptions of learning resources. This facilitates the discovery, location, evaluation and acquisition of learning resources. Sharable Content Object Reference Model (SCORM) [27] is another metadata standard. SCORM provides a reference model to develop models of learning content and delivery.

This layer also manages content development and delivery services. Authors create content, which is stored in a database. Existing content can be updated and can also be exchanged with other systems. LMS consists of complex activities such as administration, interaction among users (such as learners and trainers) through runtime environment, learner tracking, assessment etc.

Layer 4: *Data Storage and Access*—This layer provides services for data storage, data access, file access and file sharing. Typically, the services allow HTTP access, and sometimes FTP access.

Layer 5: *Infrastructure*—Services in this layer use the Internet as a communication and composition infrastructure and provides a service oriented view by using a standardized stack of protocols.

5.5 Architecture of the E-Learning Grid

In this section architecture of e-learning Grid is discussed. Grid based e-learning architecture can exploit huge computational power and data storage. Thus, through the integration of mass storage and high performance computing power, the availability and quality of service of the system also increases. A layered architecture of e-learning Grid is shown in Fig. 5.4. It contains core learning services of a LMS as well as core Grid services of Grid middleware. The learning layer services interact transparently with the Grid middleware so that a user is not aware of the Grid.

Usually the LMS coordinates all learning related activities. LMS offers both services which makes use of the Grid as well as services that does not need Grid functionality. Some of the LMS functionalities are implemented as web services. Figure 5.5 shows the detailed architecture of e-learning Grid.

Application layer—E-learning portal provides a secure unified access point in the form of a web-based user interface.

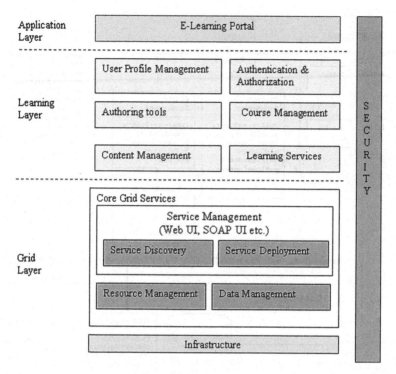

Fig. 5.4 Layered architecture of e-learning Grid

Learning layer—User management system provides functionalities like login, logout, new user creation, delete existing user etc. User management module of LMS creates a proxy certificate request for new user creation and sends it to Grid login service of Grid layer. Certificate authority (CA) of Grid signed the proxy certificate. The signed proxy certificate is sent back to LMS. This module also handles authentication and authorization mechanism of the LMS. Role of a user is used to set its access rights. For example, if a user is a learner (role) then he or she can access a learning material, the functionality of which is implemented as a course management service but denied content management service of the LMS. However, the learner can look for suitable courses, and search for learning objects in a content registry. On the other hand, the authoring tools provide an environment to create, edit and publish contents in content registry, so that they can be found by the LMS. So, users which have been assigned the role of an author can only access this service. To complete the authentication mechanism, LMS sends the login credentials to Grid login service which checks the validity of the certificate as well as its access rights. User registry is maintained and updated with user information. Figure 5.6 shows the sequence diagram of interactions between user management module of learning layer and Grid login of Grid layer.

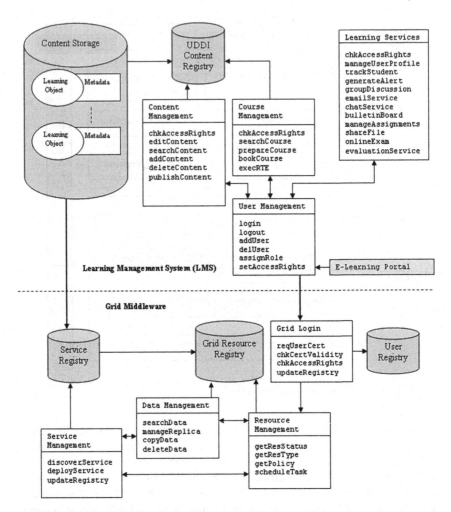

Fig. 5.5 Detailed architecture of e-learning Grid

Learning contents consist of two main components—learning object and metadata. A course material consists of several learning objects. Metadata is used to describe a learning object. The purpose of metadata is to support the reusability of learning objects, to aid discoverability, and to facilitate their interoperability. Learning objects are indexed with metadata and stored in content repository. LMS maintains its local content registry service, which uses to discover and provide learning services. The UDDI content registry is an information system that manages any content type and the standardized metadata that describe it. Content management system and course management system enables user access to the learning objects through access to WSDL files. Brief description of LMS methods are shown in Fig. 5.7.

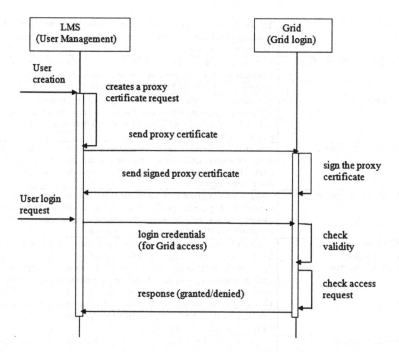

Fig. 5.6 Interactions between user management module and Grid login service

Grid layer—Some services of LMS are implemented as Grid services. Grid middleware maintains its service registry to discover and provide such services. Service management system is responsible for discovery and deployment of such services. Resource management system of Grid middleware accesses an information service which is aware of the status and type of all resources in the Grid. By accessing the Grid resource registry service it first determines which resources are available. This module is responsible for selecting the Grid resources for distributing computation and data. Data distribution and maintenance of contents of LMS is managed by data management system of Grid. LMS with Grid plays the role of service provider, i.e., exposing a set of services, helping to discover them, maintaining full control of access to them and being responsible for their execution. Figure 5.8 shows sequence diagram of content management module of LMS and Grid layer services. Description of methods in Grid middleware layer are shown in Fig. 5.9.

Web services and software agents both are a modular way to develop software. The main distinction between software agents and web services is that web services are user-driven while agents act on their own after user has given the necessary instructions. Combining web services and autonomous agents to a composite service requires common method for using the system. Web services are normally used by a e-learning user, who directs the system at each stage. Agent systems are used through client software that gives initial instructions to an agent.

login and logout	login method gathers information from user trying to log into E-Learning portal and then validate it. On success, a session is started till a logout is called.
addUser and delUser	addUser method allows creating a new user subjected to administrator's permission / approval. Administrator can delete an existing user by invoking delUser method.
assignRole	assignRole method can only be executed by administrator. This method can gather its values automatically from status of an existing user i.e. Learner / Admin / Trainer / Author unless a customization of role is sought.
setAccessRights	Administrator invokes setAccessRights method for setting access rights of a user. LMS always checks whether the user has access rights. LMS provides access and allows operations according to these access rights.
chkAccessRights	chkAccessRights is used to check the access rights. It would involve the login info (i.e. the User ID that determines the status / category of the user on which the access right actually depends) populated from User Management component.
editContent addContent deleteContent	The editContent/addContent/deleteContent method is invoked with either partial or full content edition permissibility. These methods facilitate easy inclusion and exclusion of contents as and when necessary. addContent request again should be ratified by user role (e.g., author). The base reference of the new content to be added is to be put as a child under the appropriate parent reference. In deleteContent, the respective parent reference will be linked to NULL after the operation.
searchContent	searchContent method is invoked depending on a suitable "Service event" (e.g., groupDiscussion, email or authoring) fired from the Learning Services event.
publishContent	publishContent would publish the finalized content into the respective UDDI registry. A service publishing application servlet would communicate to the registries (UDDI) using SOAP to execute the request.
searchCourse prepareCourse bookCourse	searchCourse method would allow to browse through the whole set of published courses. If a course is selected then the learner's ID would be reserved against that course/s for accountability under bookCourse method. Administrator can invoke prepareCourse method for preparing course.
execRTE	execRTE implements runtime environment service, through its communication module, receives the message, instances the run-time data and sends this instance using a SOAP message to the LMS.
generateAlert groupDiscussion emailService chatService bulletinBoard shareFile manageAssignments onlineExam evaluationService	These are session based services conducted among multiple users. At the back of each service, there should be individual application programs with distinct port numbers. Each of these services should have their individual service registries where services are described in WSDL document. Service searching or service querying applications are needed for service discovery. These applications communicate with the registries using SOAP. Service-querying applications use appropriate APIs that help in generating appropriate SOAP based queries based on UDDI specifications. Users access these services with SOAP-RPC.

Fig. 5.7 Description of LMS methods

Client software for the user is able to invoke agents, web services, or both. The agent then proceeds to do the task, reporting either success or failure. The agent may use other agents or even other software systems to do the work. These agents

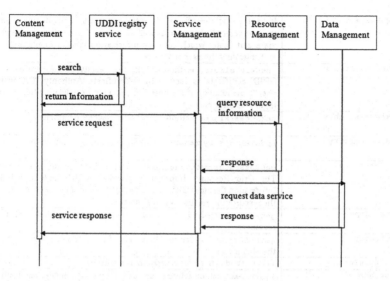

Fig. 5.8 Interactions between content management module and Grid layer services

may be as simple as web services, or they may have complex internal state and own goals. Web services may contain a hierarchy of web services.

Multi-agent system is a flexible and modular way to develop complex systems like e-learning system. Some agents do simple tasks, and the combination of multi-agent systems and web services can be helpful to reduce overhead and increase the performance of the system.

5.6 E-Learning Grid as Multi-agent System

This section presents design of the above mentioned e-learning Grid architecture as multi-agent system. The proposed framework is based on the area of e-commerce towards enhancement of e-learning services. It is already mentioned that the dynamic e-learning service has to be developed in distributed manner incorporating multi-agent techniques. In multi-agent system, multiple software agents play different kinds of roles assigned to them. In spite of having some significant responsibility, each agent can cooperate and interact with other agents to reach a common goal. With intent to achieve modular, interactive, time-oriented, context-aware services, different agents are proposed in this e-learning framework. Multiple agents along with their roles and responsibilities are elaborately discussed in the following.

In the beginning, the roles of Learning Layer and Grid Layer are identified. In the Learning Layer, the roles are User, Course Manager, Content Manager and Learning Service Manager in order to achieve the functionalities mentioned in this

reqUserCert	It creates a user certificate. A central concept in GSI authentication is the *certificate*. Every user as well as service on the Grid is identified via a certificate, which contains information vital to identifying and authenticating the user or service.
chkCertValidity	chkCertValidity method checks the validity of the certificate. A GRID certificate is issued by a Certificate Authority (CA) which checks the identity of the user. On checking or validating this, a new login session is created
chkAccessRights	chkAccessRights method checks user's access permissions.
updateRegistry	updateRegistry method is used for updating item data in the registry.
getResStatus	getResStatus method returns the Status of the resources discovered. The Grid Resource Information Service (GRIS) and Grid Index Information Service (GIIS) can be configured in a hierarchy to collect the resource information and distribute it.
getResType	getResType method returns the type and the attributes of Grid resources.
getPolicy	getPolicy method returns the policy of the resource usage verifying the SLA issues.
scheduleTask	This method is used for scheduling tasks in Grid.
searchData	searchData can be activated and driven by a searchContent method invocation form LMS
manageReplica	manageReplica method registers and manages replicas to enhance data availability.
copyData	copyData method would transact data at any user session to serve any down-level web service or web based service.
deleteData	deleteData method would delete all the data instances (replicas) present after suitable validation.
discoverService	discoverService method used for a suitable service discovery asked by any of the web service or web based service.
deployService	deployService method is responsible for deployment of services in Grid.

Fig. 5.9 Description of Grid middleware methods

layer. Specifically, the role, Content Manager involves in content management, similarly the role Learning Service Manager will be solely responsible for providing learning services of e-learning. Subsequently, in Grid layer, the role Grid Manager will perform all the service management related to the Grid middleware.

On the basis of the functionalities referred in the e-learning Grid architecture, the above-mentioned roles are subdivided into the corresponding roles as following:

1. In the User management layer, the roles Administrator, Author, Trainer and Learner will be activated with respect to the responsibilities of the role User. Administrator provides privileges for all other users and manages all the profiles. Author creates and updates contents; Trainer moderates contents

(prepared by more than one Author), prepare assignments, evaluate assignments and also involves in synchronous learning services like, chat and asynchronous services like, email.

2. In the Learning Management Service (LMS) layer, corresponding roles are same as aforementioned roles of Learning Layer.
3. In the Grid middleware layer, the role Grid Manager acts as collaborator between LMS layer and Grid layer. More specifically, Grid Manager needs to be further categorized into Grid Resource Manager, Grid Data Manager and Grid Service Manager. Grid Resource Manager allocates resources for execution of task as and when required, Grid Data Manager is responsible of distribution and maintenance of data. Grid Service Manager is responsible of discovery and deployment of services in Grid.

Hence, the detailed roles of each layer are mapped into the consequent agents and agent mapping is shown in Fig. 5.10. This agent model would definitely provide a direction towards implementation of the system.

In addition to this agent model, it is very much necessary to discuss role, responsibility and granted permission of each agent for each layer distinctively. At first, User Management Layer, all these four agents: Administrator Agent, Author Agent, Trainer Agent and Learner Agent are distinguishable through their access rights only. Each of these user enters into the system through portal login itself. Any individual can play multiple roles, but not through single login. Purposely, the users are specialized into these four agents. Administrator Agent is solely responsible for assignment of different roles and access rights to the users according to their requirements. It creates new user, deletes existing user if not using the system for long time or sent some requests for not willingness to use. Most importantly, it manages user profile. Simultaneously Administrator Agent

Layers	Basic role	Detailed role	Agent
User management	User	Administrator	Administrator Agent
		Author	Author Agent
		Trainer	Trainer Agent
		Learner	Learner Agent
Learning Management Service	Course Manager	Course Manager	Course Agent
	Content Manager	Content Manager	Content Agent
	Learning Service Manager	Learning Service Manager	Learning Service Agent
Grid middleware	Grid Manager	Grid Manager	Grid Agent
		Grid Resource Manager	Grid Resource Agent
		Grid Data Manager	Grid Data Agent
		Grid Service Manager	Grid Service Agent

Fig. 5.10 Mapping of roles and agents in different layers

associates with other four agents. Author Agent is actually the writer of the book, document, course material or content of this e-learning Grid. So it can create the new content, updates or deletes the existing contents. Thus, the responsibility of Author Agent is publishing the content or learning materials in content registry for availability of learning. Next, Trainer Agent performs interaction between different learners through asynchronous or synchronous learning services. Trainer Agent can use the existing available contents or they can moderate the contents according to the specific course. It involves in preparation of specific assignments and evaluation as well. Learner Agent plays the most important role in this e-learning Grid with respect to utility. This agent searches for specific courses and accesses course based learning materials on the basis of availability of the courses. It involves in synchronous or asynchronous interaction with trainer as doubt clearing session or improvement of the learning courses.

Next, in Learning Management Service Layer, checking of access rights is essential for the agents namely, Course Agent, Content Agent and Learning Service Agent. Course Agent is exclusively in charge of managing all available or new courses according to the learners choice. First of all, it reserves the specific courses or lessons for each user easy access and then it also checks the requirements and finally can set up a new course if necessary. Partly, contents of the courses or lessons are managed by Content Agent. Content Agent accesses the available contents which are published by the Author Agent in the content registry. It is responsible for preparing study material of the specific courses while the parts of the contents (created by the Author Agent) can be used in more than one course simultaneously. However, sometime same contents can be used in various courses partly. Thus, the Content Agent can often interact with the Author Agent. Subsequently, Learning Service Agent provides diverse e-learning services to the learners. Learning services may be synchronous or asynchronous based on the courses. This agent is part of the Course Agent as learners can access the different services through this agent. These agents are solely responsible for conducting online examinations, appraises the learners after evaluating the assessments. At the same time, it can also generate alerts to the learners through continuous tracking. At this stage, it is obvious that Learner Agent and Trainer Agent are dependent on Course Agent. Because Course Agent makes available the courses or lessons to the Learner Agent and Trainer Agent just does interaction with the learners for that specific courses.

At last, in Grid Middleware Layer, responsibility of Grid Agent acts as service provider in Grid. Interaction between Grid Agent and user of e-learning portal is transparent to the user but it is indeed required. Therefore, each user is dependent on Grid Agent for providing heterogeneous resources, enormous data and reliable services. Except checking access right, it reads registered users certificates and validates of these certificates. Entirely it exposes the services through updating registry; maintains users accessibility. Since Grid Agent plays the role of service provider, it is sub divided into special category of agents, Grid Resource Agent, Grid Data Agent and Grid Service Agent depending upon their responsibilities. These three agents interact themselves too. Responsibility of Grid Resource Agent

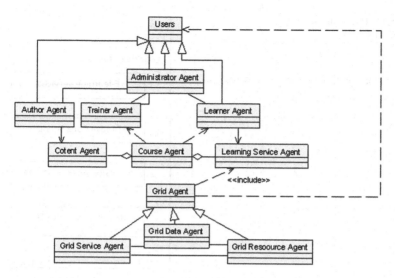

Fig. 5.11 Agent class diagram

is allocation of resources for execution of task as and when required. In order to do this task, first it checks access rights, and then reads user policy along with resource type. Finally, it schedules the specified task after checking the resource status. Distribution and maintenance of data in Grid is carried out by Grid Data Agent. Obviously, after checking access rights, it searches specific data according to learner need then makes available the data through copying it; can delete data if it is obsolete; finally, can manage replicas for easy retrieval and reliability. Grid Service Agent is solely responsible for discovery and deployment of services in Grid. So as to make available the services, it also updates service registry. Interactions of these agents are represented in the class diagram (Fig. 5.11).

5.7 Implementation of E-Learning Service with Globus Toolkit

This section discusses the implementation details of e-learning service with Globus toolkit, version 4 (GT4). GT4 is builds upon web services standards and technologies like Simple Object Access Protocol (SOAP) and Web Services Description Language (WSDL). The main components of the server side of GT4 is shown in Fig. 5.12. The GT4 architecture consists of a Grid container to manage all of the deployed Grid services throughout their lifecycles. The GT4 Grid container uses apache axis [28] as its Grid services engine to handle all of the SOAP message processing, JAX-RPC handler processing, and services configuration. GT4 provides software libraries that support security, discovery, resource

Fig. 5.12 E-learning service
as a Grid service

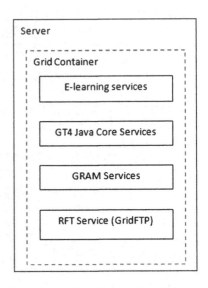

management, invocation, communication, exception handling, data management, etc. Programming models for exposing and accessing service implementations such as Grid Resource Allocation Management (GRAM) and GridFTP are also supported by GT4. The Reliable File Transfer Service (RFT) is a web service that provides interfaces for controlling and monitoring third party file transfers using GridFTP servers. GRAM uses RFT for staging operations and RFT uses GridFTP to perform the actual transfer of data. E-learning services use these services to transferring files from one machine to another. GT4 Java core services offer a run-time environment capable of hosting user application services. The run-time environment mediates between the user-defined application services and the GT4 core services, underlying network, and transport protocol engines.

E-learning system is a collection of several services and these services are implemented as Grid services. To create and deploy a service of e-learning system as a Grid service, we need to do the followings. Here we consider content management service of e-learning system as an example.

- *Service interface*: A WSDL file is created to define the e-learning service's interface. The WSDL file contains the abstract definition of the service including the types, messages and portTypes. For example, the portTypes for the e-learning content management service are defined in a WSDL file (content.wsdl). It describes all operations that the content management service will provide. A portType defines one or more operations using the operation element. Each unique operation element defines an operation and the input/output messages associated with the operation. The operation elements within a portType define the syntax for calling all methods in the portType. The content data type is defined in a xsd file (content.xsd). Messages are also defined here. A Message element consists of one or more part sections. Each part element corresponds to a parameter. Each part has a type attribute. Messages can be either Requests (input messages) or Responses (output messages). In this step the service endpoint interface (e.g., ContentMgmtPortType) is generated.

- *Service implementation*: In this step, a class is implemented the interface defined in the previous step. For example, here we provide a class, ContentMgmtImpl that implements the ContentMgmtPortType interface. The implementation uses stub classes that were generated from the WSDL file. The ContentMgmtImpl class implements the methods defined in the ContentMgmtPortType. In this way all the services of e-learning system are implemented.

- *Deployment parameters*: A Web Services Deployment Descriptor (WSDD) file is created to define the deployment parameters. The GT4 deployment descriptor is the deploy-server.wsdd. This file contains a service name which specifies the location of the service.

- *GAR file generation*: In this step all source codes are compiled and a GAR file is generated using Ant. The GAR file contains all the files and information that the web server needs to deploy the e-learning service.

- *Service deployment*: In this step the GAR file is deployed using tools from the GT4 distribution. Gt4 deployment tool globus-deploy-gar is called to copy the archive files (wsdl file, compiled stubs, compiled implementation, wsdd file) into the appropriate server location directory tree of the GT4 container.

Consequently, in order to serve these above-mentioned services (Sects. 5.4 and 5.5) autonomously or through directed by any other service in e-learning Grid, java-based client or user side (namely, Learner Agent, Trainer Agent, etc.) and server or Grid side (like, Grid Resource Agent, Grid Data Agent, etc.) agents would also be deployed.

5.8 Benefits of E-Learning Grid

Traditional e-learning systems are often based on technologies which are difficult to scale-up or share with multiple users. Moreover, performance of e-learning systems degrades for data intensive fields, such as image processing, multimedia etc. In this section, we analyse how the proposed e-learning system utilizes underlying Grid environment to tackle these problems.

Grid environment help learners to worldwide collaborate and carry out effective learning. Grids allow users to create virtual learning communities and organizations. These virtual organizations are providing a platform to share learning resources, ideas and views. This also makes the system scalable to a potentially unlimited number of parties. On the other hand, one of the popular ways to make educational content available is by recording lectures. However, the main drawback about recording lectures is that the better the quality of the recording, the larger the file that stores it. Thus, huge capacity storage devices are required to manage a single course which may comprise several hours of recorded lectures with reasonable video and audio quality. Moreover, sharing resources would be of particular benefit, where learners use portable devices with limited memory and processing power. Data intensive fields, which require significant amounts of computing power, can easily get benefit from underlying Grid environment of e-learning Grid.

By sharing the processing power and storage space of resources of Grid environment, the proposed system could offer a viable solution. More specifically, the main ways that Grid technologies can leverage educational activities in the proposed e-learning Grid system are as follows:

- *Virtual organization*: The users of e-learning system can be organized dynamically into a number of virtual organizations. For example, creating virtual classrooms by interconnecting trainers to geographically scattered learners. Within virtual organizations, users can access any resources without knowing the exact location of the resources. This includes real-time and asynchronous collaboration. Moreover, educators and students can access a massive computing power for simulations and computational calculations which is a problem in traditional e-learning systems.
- *Data-Grid*: Learning material plays a major role in e-learning; this includes recorded video lectures, tutorials, books, articles and so on. This requires huge storage capacity that may be easily scaled-up whenever required. Grid environment accomplishes the problem of storing huge quantities of data, which demands high storage capacity infrastructures.

5.9 Conclusion

The main intention of this chapter is to draw up a research agenda for an exploitation of Grid computing in the field of e-learning. This chapter presented an overview of the e-learning Grid environment and discussed different services provided by the e-learning Grid middleware. In order to provide dynamic e-learning service, it needs to be developed as distributed application incorporating multi-agent techniques. E-learning services need to be developed scalable, flexible, secured as it is necessary to be accessed by users located at sites distributed over geographically. Grid middleware provides functionalities to manage resources in the dynamic, distributed, heterogeneous environment. Multi-agent systems offer valuable techniques to provide the autonomy and flexibility required in highly dynamic and heterogeneous environments. These are also defining characteristics of Grid environments. E-learning application would benefit from incorporating such techniques. Design of e-learning Grid framework is presented here as multi-agent system. Architecture of e-learning Grid which integrates core Grid middleware and LMS functionality appropriately is also outlined in detail. Implementation of a service of e-learning system as a Grid service is discussed. In future, we intend to extend the implementation of e-learning Grid with GT5 (latest version).

References

1. Foster, I., Jennings, N.R., Kesselman, C.: Brain meets brawn: why Grid and agents need each other. Published in 3rd International Conference on Autonomous Agents and Multi-Agent Systems, New York, USA, pp. 8–15 (2004)
2. Wang, K., Ke, J., El Saddik, A.: Architecture for personalized collaborative e-learning environment. In: Proceedings of World Conference on Educational Multimedia, Hypermedia and Telecommunications, pp. 4801–4805 (2005)
3. Casella, G., Costagliola, G., Ferrucci, F., Polese, G., Scanniello, G.: A SCORM thin client architecture for e-learning systems based on web services. Int. J. Distance Educ. Technol. 5(1) (2007)
4. Su, M.T., Wong, C.S., Soo, C.F., Ooi, C.T., Sow, S.L.: Service-oriented e-learning system. In: First IEEE International Symposium on Information Technologies and Applications in Education (ISITAE'07), pp. 6–11 (2007)
5. Gaeta, A., Gaeta, M., Ritrovato, P.: A Grid based software architecture for delivery of adaptive and personalised learning experiences. Pers. Ubiquit. Comput. 13(3), 207–217 (2009)
6. Li, Y., Yang, S., Jiang, J., Meilin, S.: Build Grid-enabled large-scale collaboration environment in e-learning Grid. Expert Syst. Appl. 31(4), 742–754 (2006)
7. AlZahrani, S., Ayesh, A., Zedan, H.: Mult-agent based dynamic e-learning environment. Int. J. Inf. Technol. Web Eng. 4(2) (2009)
8. Andreev, R.D, Troyanova, N.V.: E-learning design: an integrated agent-Grid service architecture. In The Proceeding of IEEE John Vincent Atanasoff 2006 International Symposium on Modern Computing (JVA'06) (2006)
9. Kashfi, H., Razzazi, M.R.: A distributed service oriented e-learning environment based on Grid technology. In: The Proceeding of 18th National Computer Conference (2006)
10. Shen, Z.Q., Gay, R., Miao, C.Y., Wang, Q.: Goal oriented modeling agent mediated e-learning Grid. Control, Automation, Robotics and Vision, Kunming, China, 6–9 Dec 2004
11. Foster, I., Kesselman, C., Tuecke, S.: The anatomy of the Grid: enabling scalable virtual organizations. Int. J. High Perform. Comput. Appl. 15(3), 200–222 (2001). (Sage Publishers, London)
12. Joseph, J., Ernest, M., Fellenstein, C.: Evolution of Grid computing architecture and Grid adoption models. IBM Syst. J. 43(4), 624–645 (2004)
13. Globus: web site http://www.globus.org
14. Ferreira, L., Berstis, V., Armstrong, J., Kendzierski, M., Neukoetter, A., Takagi, M., Bing-Wo, R., Amir, A., Murakawa, R., Hernandez, O., Magowan, J., Bieberstein, N.: Introduction to Grid Computing with Globus. IBM Redbooks. Available from: http://www.redbooks.ibm.com/redbooks/pdfs/sg246895.pdf
15. Global Grid Forum: web site http://www.ggf.org/
16. Tuecke, S., Czajkowski, K., Foster, I., Frey, J., Graham, S., Kesselman, C., Maguire, T., Sandholm, T., Snelling, D., Vanderbilt, P.: Open Grid Services Infrastructure (OGSI) Version 1.0. Available from: http://www.globus.org/toolkit/draft-ggf-ogsi-gridservice-33_20030627.pdf (2003)
17. OASIS Web Service Resource Framework (WSRF): http://www.oasis-open.org
18. Foster, I., Czajkowski, K., Ferguson, D., Frey, J., Graham, S., Maguire, T., Snelling, D., Tuecke, S.: Modeling and managing state in distributed systems: the role of OGSI and WSRF. Proc. IEEE 93(3), 604–612 (2005)
19. Foster, I., Frey, J., Graham, S., Tuecke, S., Czajkowski, K., Ferguson, D., Leymann, F., Nally, M., Storey, T., Weerawaranna, S.: Modeling Stateful Resources with Web Services. Globus Alliance. Available from: http://www.ibm.com/developerworks/library/ws-resource/ws-modelingresources.pdf (2004)

20. Burbeck, S.: The Tao of e-business Services: The evolution of Web applications into service-oriented components with Web services. Emerging Technologies, IBM Software Group. Available from: http://www.ibm.com/developerworks/webservices/library/ws-tao (2000)
21. Srinivasan, L., Treadwell, J.: An overview of service-oriented architecture, web services and Grid computing. HP Softw. Global Bus. Unit. Available from: http://devresource.hp.com/drc/technical_papers/grid_soa/index.jsp (2005)
22. Mei, Q., Shen, J.: A knowledge flow driven e-learning architecture design: What is its stratification and how is it personalized, computers in education. In: International Conference on Computers in Education (ICCE'02), pp. 1307–1308 (2002)
23. Nejdl, W., et al.: EDUTELLA: A P2P networking infrastructure based on RDF. In: World Wide Web Conference Series (WWW2002), Honolulu, Hawaii, 7–11 May 2002
24. Shen, Z., Shi, Y., Xu, G.: A learning resource metadata management system based on LOM specification. In: Proceedings of the 7th International Conference on Computer Supported Cooperative Work in Design, pp. 452–457, 25–27 Sept 2002
25. Li, H.T., Lin, C.H., Chang, Y.C., Yang, J.T.D: On the distributed management of SCORM-compliant course contents. In: Proceedings of IEEE International Conference on e-Technology, e-Commerce and e-Service, pp. 538–539 (2004)
26. Apache Axis: web site http://axis.apache.org/axis/
27. Foster I., C. Kesselman, J. M. Nick and S. Tuecke. The Philosophy of the Grid: An Open Grid Services Architecture for Distributed systems Integration. Open Grid Services Infrastructure WG, Global Grid Forum. Available from: http://www.globus.org/alliance/publications/papers/ogsa.pdf (2002)
28. Germanakos, P., Tsianosl, N., Lekkas, Z., Mourlas, C., Belk, M., Samaras, G.: A semantic approach of an adaptive and personalized web-based learning content—the case of adaptive web. In: Second International Workshop on Semantic Media Adaptation and Personalization (2007)

Chapter 6
Determining the Usability Effect of Pedagogical Interface Agents on Adult Computer Literacy Training

Ntima Mabanza and Lizette de Wet

Abstract A large part of the population in developing countries is technologically ignorant. Pedagogical interface agents are pieces of educational software with human characteristics that facilitate social learning. The aim of this research was an attempt to evaluate the extent to which a variety of pedagogical educational agents could assist adult learners in acquiring basic computer skills. This was done by conducting a usability test in the context of South African adult computer literacy training. A hundred and three participants were randomly assigned to either a control group or a test group, where after all participants received Microsoft Office Word training (pre-test). Only test group participants were introduced to pedagogical agents (experimental treatment). During the usability test both groups were given tasks to perform. Findings showed that computer illiterate adult users could perform better during literacy training with the assistance of educational agents when compared to only being taught through traditional teaching methods. This could open the doors to more effective ways of reaching and teaching a larger group of previously educationally disadvantaged adults in order to give them a better chance at securing employment in the labour market.

Keywords Pedagogical agents · Educational agents · Adult learners · Usability testing · Computer literacy

N. Mabanza (✉)
Department of Information Technology, Central University of Technology Free State,
Bloemfontein, South Africa
e-mail: nmabanza@cut.ac.za

L. de Wet
Department of Computer Science and Informatics, University of the Free State,
Bloemfontein, South Africa
e-mail: lizette@ufs.ac.za

M. Ivanović and L. C. Jain (eds.), *E-Learning Paradigms and Applications*,
Studies in Computational Intelligence 528, DOI: 10.1007/978-3-642-41965-2_6,
© Springer-Verlag Berlin Heidelberg 2014

6.1 Introduction

Similar to other developing third world countries, South Africa (SA) is also challenged by adult illiteracy. The term 'adult illiterate' refers to a person with little or no formal education. According to Professor Solomon Sibiya of the University of Pretoria, illiteracy among the population in SA is preventing young people and adults from effectively participating in the social, economic and political life in the new SA [1].

Illiteracy is but one of the factors contributing to the high level of computer illiteracy and technological ignorance among the population of SA. Other factors include lack of funds, lack of infrastructure, shortage of computer instructors, and emotional factors such as fear of the unknown. Computers have become part of our lives. Various jobs require the use of computers as part of everyday tasks. There is a need to find better ways to support computer illiterate people in SA so that they can become part of the workforce, raise their self-confidence and feelings of worthiness and also enabling them to take part in labour and social activities. This can be achieved by giving them the necessary learning opportunities such as basic computer training that can improve their skills development for employability.

For several years now, among Human–Computer Interaction (HCI) researchers, there have been increased efforts towards developing innovative tools to ease or enhance users' interaction with computers. Pedagogical Interface Agents (PIAs) is one example of such an innovative tool. A PIA is a piece of educational software with human characteristics used for the purpose of assisting a learner in the completion of his/her tasks in a socially engaging manner [2]. The main motive behind these innovative tools is to reduce novice users' perception of the learning difficulty level of the material, as well as to help them in managing or recovering from negative emotions that might arise during their interactions with computers. For these reasons it was decided to pursue the use of PIAs in the research project discussed in this chapter.

In line with this, the research project in question was seen as an initial step towards finding ways to incorporate PIAs to ease computer literacy training for adult computer illiterate users in SA. This research was conducted with the co-operation of a group of adult learners from the Mangaung University of the Free State Community Capacity Programme (MUCCP) based in Bloemfontein, SA. These adults had little or no formal post-school education and exposure to computers. They were introduced, trained, and assessed using a Simulated Microsoft Office System (SMOS) developed by Potgieter [3] at the University of the Free State. This simulated system incorporated a variety of PIAs (varying in terms of e.g. appearance, gender, voice, and reality).

Usability is the key to any successful software system and can broadly be explained as how effective, efficient and enjoyable (satisfying) a system is to use [4]. The main focus of this research study was to test the usability of a variety of PIAs incorporated in an SMOS to determine the extent to which each of them could help adult computer illiterates in acquiring basic computer skills. To accomplish this, test participants' opinions (satisfaction) on each of these PIAs (in terms of e.g. appearance, voice and movement), as well as their performance while making use of the assistance provided by the PIAs when carrying out basic Microsoft (Ms) Office Word tasks, were explored.

The remainder of this chapter is structured as follows: Sect. 6.2 introduces the main research areas, terms and terminology relevant to this study. It further discusses the current research objectives and approaches used in this research, its limitations, and its contributions. Section 6.3 discusses related work on pedagogical agents and agent systems. It also explains the Simulated Microsoft Office System (SMOS) and the various kinds of agents incorporated in this system. Section 6.4 explains the design approaches and methodology used to carry out this study. In Sect. 6.5 the analysis and interpretation of data collected is presented. The chapter concludes with Sect. 6.6 where the study findings, lessons learnt and possible future research is also addressed.

6.2 Background

In order to fully grasp the scope of this research study, it is important to first clarify the main terms and relevant terminology. Thereafter the research objectives and approach will be outlined, followed by the limitations of the study, as well as its main contributions.

6.2.1 Terms and Terminology

Interface Agents. There are several suggested definitions of what an interface agent is. According to Rudowsky [5], Giraffa and Viccari [6] there is no universally accepted definition of the term in the research community. Rudowsky [5] stated that there is consensus that autonomy (the ability to act without the intervention of humans or other systems) is a key feature of an interface agent. Giraffa and Viccari [6] added that agents must have the following properties: reactiveness, autonomous, goal driven or utility driven, temporally, continuous, mobile, flexible, and representing a character. Several interchangeable terms are used in the research community when referring to interface agents. They include: agent, user interface agent, intelligent agent, software agent, and emotional interface agent.

Although there is no universal accepted definition of the term interface agent, for the purposes of this study, the researchers opted for the definition of Lincicum [2] who described an interface agent as a character enacted by a computer that interacts with the user in a socially engaging manner.

From the definition above it can be inferred that the main purpose of an interface agent should be to ease the human-technology interaction, as well as to make the use of software a more enjoyable experience. To achieve this purpose, interface agents must be given anthropomorphic characteristics. According to Bartneck and Kulic [7] anthropomorphism refers to the attribution of human form, human characteristics, or human behaviour to non-human things such as robots, computers, and animals.

Interface agents are often represented by anthropomorphic bodies [8]. Their most important utility is to facilitate the interaction with the user in natural language, through voice recognition, or via textual input and output. Thus, an anthropomorphic interface agent can be considered as a character that is life-like (often human or animal), able to exhibit realistic human movements (such as talking, walking, and running), and having distinct personalities.

Four main types of interface agents were identified, namely contextual, non-contextual, metaphoric, and abstract agents [9]. From these four, the contextual agent will be the focus for this study.

Interface agents are used or implemented in different disciplines and application domains. These include e-Commerce, entertainment, medicine, and education. The application of interface agents in education (then referred to as educational agents) will be the core of this research study.

Educational Agents. Educational agents are pieces of educational software with human characteristics that facilitate social learning. The characteristics of the agent can be expressed to students in text, graphics, icons, voice, animation, multimedia, or virtual reality [10].

A number of different types of educational agents exist. However, a classification of agents provided by Chou et al. [10] divided educational agents into two major categories, namely personal assistants and pedagogical interface agents. These two types of educational interface agents can be designed to perform a human instructional role by doing diverse tasks. A personal assistant can perform as a teacher assistant or a learner assistant, while a pedagogical agent can perform as a tutor or a co-learner [11]. A tutor agent plays the role of a teacher, while a co-learner agent plays the role of a learning companion.

For the purpose of this research, pedagogical interface agents will be the focus, especially agents who play the role of tutors.

Furthermore, the emphasis in terms of the population will be on adult computer illiterate users and therefore also on the use of pedagogical interface agents in adult computer literacy training.

Adult Computer Literacy Training. The term 'computer literate' refers to a person who has some basic knowledge of the use of computers [12]. In relation to this definition, one can say that the key objective of adult computer literacy training is to assist adult learners to acquire basic knowledge and necessary skills to use computers to perform basic tasks. In SA, to date, adult computer literacy is generally conducted using traditional educational training approaches (i.e. conducted by a skilled human instructor). However, each individual learns at a different pace. Often this traditional training situation poses intellectual challenges to certain adult learners, reasons being certain factors such as lack of self-esteem and confidence mostly due to the lack of basic education at a younger age. Education providers need to develop innovative solutions for reaching larger numbers of those less familiar with technology, such as poor, less educated (both young and older) adults [13], otherwise these individuals will become further marginalized in our society.

Although the adult learners who were used as participants in this study were literate in terms of being able to write and read, they had little or no previous knowledge of computers and also had limited learning experiences. Their ages varied, with some not having had access to learning material for many years. It is to be expected that computer literacy training that involves user-computer interaction with a mouse, keyboard, typing, etc. as necessities, presents another set of challenges for these adult learners to be anxious or nervous about. Therefore, these facts were taken into consideration when deciding on how the adult computer literacy training session would best be conducted. The big challenge was finding the best approach for simplifying the computer literacy training for all these adult learners without compromising the quality of training.

In order to determine the best approach, best techniques and best aesthetics in software applications, usability evaluation to investigate usability is a necessity. In this study, the incorporation of various kinds of pedagogical interface agents during computer literacy training needed to be evaluated.

Usability. There are many proposed definitions for usability. A few of the well-known definitions are:

- Usability is a quality attribute used to assess how easy user interfaces are to use [14]
- Usability refers to how well users can learn and use a product to achieve their goals and how satisfied they are with that process [15]
- The effectiveness, efficiency and satisfaction with which specified users achieve specified goals in particular environments [4].

In the context of this study, the proposed ISO definition [4] as stated above, was adopted; this definition highlighted three important notions of usability, namely effectiveness, efficiency and satisfaction. Hence the use of these three notions as the focal point for the usability tests conducted in this research.

Usability evaluation is conducted using a variety of techniques, usability testing being one of them. Usability testing is a formal approach which is usually conducted in a controlled laboratory environment. The goal of usability testing is to identify any usability problems, collect quantitative data on participants' performance (e.g. time on task, error rates), as well as to determine user satisfaction [16]. It involves measuring the performance of typical end-users as they undertake a predefined set of tasks on the system being evaluated to assess the degree to which it meets specific usability criteria [17].

In this study, the uniqueness lies in the fact that the major aim of this usability measurement (as will be discussed in more detailed in the following section) was to allow the researchers to assess the level of ease of use of a variety of pedagogical interface agents by adult learners to execute their basic Ms Office Word tasks effectively, efficiently and with satisfaction while becoming computer literate at the same time.

6.2.2 Research Objectives and Approaches

Based on the major aim mentioned in the previous section, the following research objectives were set:

- To measure the usability effect of incorporating pedagogical interface agents on adult computer literacy training.
- To describe the changes in knowledge, attitude and aspiration of computer literacy training program participants based on their computer literacy training with pedagogical interface agents.
- To suggest ways in which to incorporate pedagogical interface agents in order to improve adult computer literacy training.

With these three research objectives in mind, the following research question was pursued:

What is the usability effect of pedagogical interface agents on adult computer literacy training?

The research question led to the formulation of three hypotheses:

$H_{0,1}$ There is no difference in the usability performance in terms of efficiency when using a pedagogical interface agent to train adult computer illiterates when compared to using traditional educational training techniques.

$H_{0,2}$ There is no difference in the usability performance in terms of effectiveness when using a pedagogical interface agent to train adult computer illiterates when compared to using traditional educational training techniques.

$H_{0,3}$ There is no difference in the user satisfaction when using pedagogical interface agents to train adult computer illiterates when compared to using traditional educational training techniques.

A usability test was conducted in order to test the three hypotheses stated above. Usability performance and satisfaction were the two types of usability metrics captured during the usability testing. Usability performance was represented by effectiveness and efficiency, whereas user satisfaction was captured from users' opinions about the pedagogical interface agents used in this study.

6.2.3 Research Limitations and Contributions

The following can be seen as limitations of this research project:

- Complex tasks were not included.
- Not all the aspects of efficiency were taken into consideration (i.e. time taken to complete the tasks and number of steps required to complete a task).

As far as the contribution of this study is concerned, it throws light on how pedagogical interface agents can enhance computer literacy training in a third world context (in terms of better quality training and reaching more people simultaneously without the restriction of not having enough human instructors), where being trained to become computer literate could throw a life line to many adults with very little or no post-school education to give them a better chance at finding a decent job with career growth prospects.

6.3 Related Work

One should not re-invent the wheel. Additionally, it is never harmful to learn from other people's mistakes or gain from their triumphs. Therefore, literature related to the research project that forms the major topic of discussion in this chapter should be investigated briefly. This will be done by firstly looking at studies on pedagogical agents, followed by a few examples of pedagogical agent systems already in use.

6.3.1 Studies on Pedagogical Interface Agents

Most of the research done in the field of pedagogical interface agents is commonly characterized by empirical evaluation studies mainly focussing on the different factors affecting agent interactions, such as agent realism, characteristics of users (e.g. gender and age), as well as characteristics of the task.

In terms of agent realism, various studies have investigated the level in which the agent resembles a real person or animal in form or behaviour by comparing the outer appearances of the agent to a real person or animal. For example, researchers looked at the effect of agent realism on users' perception [18, 19]. In other studies the impact of agent realism on the users' achievement was measured [20, 21], as well as how physiological resemblance between agent and user influences a user's choice of agent [22, 23].

In addition to realism, there have also been various studies with regard to user characteristics. In these studies various features of the user such as gender, age, and psychological states have been investigated. For instance, White et al. [24] described an experiment designed to measure whether the non-verbal behaviour of the agent made a difference to users' interactions with the system or not. Kim and Wei [25] investigated the impact of learners' attributes on their choice of pedagogical agent used in the learning environment.

The majority of available studies on pedagogical interface agents have focused on childhood to undergraduate, college-aged adult populations [26]. Hence the incorporation of pedagogical interface agents into adult learning has thus far been largely left unexplored.

Similar to previous studies, this study also investigated some of the different factors affecting user-agent interactions, such as the agent's appearance, voice, movement, gender, etc.

6.3.2 Pedagogical Interface Agent Systems

Researchers have for several years investigated the potential of pedagogical interface agents to promote learning. As a result, there are many virtual human agents in simulated real world environments that have been developed in research laboratories by different research groups or by individual researchers around the world.

Soar Training Expert for Virtual Environments (Steve), and Agent for Distributed Learning Environment (Adele) are some examples of pedagogical interface agent systems developed by the Centre for Advanced Research in Technology for Education (CARTE) at the University of Southern California (USC).

Steve. Steve [27], shown in Fig. 6.1, is an agent that provides both individual and personal training on how to operate engines, and helps and provides guidance aboard US Navy surface ships.

Adele. Adele [28] supports online case problem solving for medical students, particularly in the family medicine and graduate level geriatric dentistry. It monitors them as they work through a simulated case. Adele is shown in Fig. 6.2.

The Computer Virus Educational System, Dr Evan, Expert Agent (EXA), and Mentor Agent (MEA) are examples of systems that were inspired by some of the earlier research mentioned above and will now be discussed briefly.

The Computer Virus Educational System. This system [29] is a tutoring system that teaches students a range of subjects on computer virus prevention techniques. The Computer Virus Educational System is shown in Fig. 6.3.

Dr. Evan. Dr. Evan [30] is a virtual health inspector agent based on simulation for teaching and training hospital workers on the proper procedures of hand hygiene.

Expert Agent (EXA) and the Mentor Agent (MEA). *EXA* and *MEA* are two versions of agents with different cognitive styles, in multimedia learning

Fig. 6.1 Steve [27]

Fig. 6.2 Adele [28]

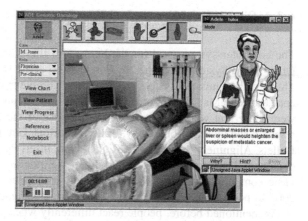

Fig. 6.3 The computer virus educational system [29]

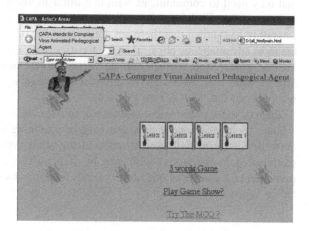

environments, developed to instruct undergraduate students [31]. These two versions of pedagogical interface agents have different instructional roles. The EXA instructional role is to give the information directly when questions are asked by students, while MEA's instructional role is to provide sufficient guidance for students to search for information.

Literature has revealed that pedagogical interface agents have been developed for various applications ranging from training to education. With regard to the existing pedagogical agents, the literature further revealed that pedagogical interface agents were usually not tested in a learning office environment. Additionally, research done in the past did not focus on adult computer illiterates [3]. The novelty of this study, therefore, lies in the fact that it focuses only on adult learners, in particular computer illiterate adults in a third world country.

The Simulated Microsoft Office System (SMOS) that was used in this study is discussed in detail in Sect. 6.3.3.

6.3.3 Simulated Microsoft Office System

The computer system used for the purpose of this research was a Simulated Ms Office Word System developed by Potgieter [3] at the University of the Free State. This was a simulated system that ran on a computer desktop. It was similar in appearance to Ms Office Word 2007 but had limited functionality. These functions were those that were necessary for the purpose of the training conducted in this research and include Bold, Copy, Paste, Underline, Cut, Italic, Font (Size, Colour), Bullets, Alignment (Left, Centre, Right), Find, Insert Picture, New document, Save document, Undo and Redo. The purpose of the training was to teach adult learners basic computer literacy skills which could enable them to use Ms Office Word 2007 to perform basic office tasks (i.e. create a document, type CVs, etc.).

As mentioned before, people react differently when confronted with characters that they need to communicate with that differ in visual appearance. This implies that they relate to them differently on an emotional, as well as a physical level. Literature has shown that there are quite a number of different opinions regarding the most advantageous physical characteristics of an agent. Aspects that may differ include gender, age, realism, human versus animal, and voice versus text. These aspects were used to design 10 different pedagogical interface agents that were incorporated into SMOS (see Fig. 6.4).

Each of these pedagogical interface agents shown in Fig. 6.4 played the role of a tutor. Their assistance mainly consisted of providing the participants (from here onwards this term will used to refer to the adult computer illiterate users that participated in the study) with step by step instructions on how to perform a particular task using any one of the various functions included in the simulated Ms Office Word environment.

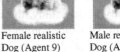

Female text agent (Agent 1) Male text agent (Agent 2) Female text & audio (Agent 3) Male text & audio (Agent 4) Female cartoon human (Agent 5)

Male cartoon human (Agent 6) Female cartoon Dog (Agent 7) Male cartoon Dog (Agent 8) Female realistic Dog (Agent 9) Male realistic Dog (Agent 10)

Fig. 6.4 Ten pedagogical interface agents incorporated in SMOS [3]

Fig. 6.5 Choose an agent
screen

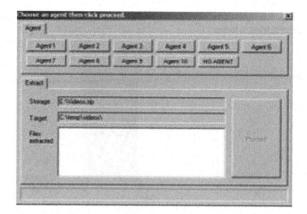

SMOS consisted of two main components, namely 'choose an agent' and 'launch office-agent application'. The function of each of these system components will now be explained briefly.

Choose an agent. In order to interact with the system, a participant first had to select the kind of pedagogical interface agent that he/she wanted to interact with. This choice was made on the 'Choose agent' screen (Fig. 6.5). The following are the various steps that had to be followed in order to select a particular pedagogical interface agent:

- A double click on the 'Choose agent' icon on the desktop opened the 'Choose an agent' screen.
- The 'Choose an agent' screen contained a list of buttons labelled Agent 1, Agent 2, etc. up until Agent 10. Each of these labelled buttons represented one of the ten pedagogical interface agents (see Fig. 6.4). The participant had to click on one of those buttons indicating which agent's assistance was needed.
- A click on the 'Proceed' button extracted the selected agent.

After selecting a particular pedagogical interface agent, the next step involved the launching the SMOS application.

Launch SMOS. In order to start the SMOS application to get help from an agent, the participant had to double click on the Ms Office Simulation shortcut icon on the desktop. By doing so, an Ms Office Simulation window (Fig. 6.6) was opened that appeared similar to Ms Office Word 2007.

The following steps summarize how a participant could get an agent's help on a particular function:

- In the Ms Office Word 2007 Simulation Window, the participant had to click on the 'Help' drop-down-list which was located next to the 'View' tab on the ribbon.
- A pull-down list appeared containing a list of Ms Office Word functions (e.g. bold, cut, exit, etc.) included in the system.

Fig. 6.6 Ms office word
simulation window

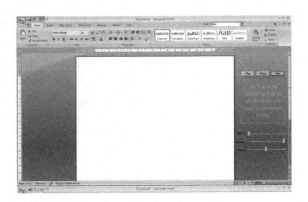

- From the pull-down list, a participant could select a particular Ms Office Word function that he/she needed assistance in.
- After choosing the particular function, the pedagogical interface agent (e.g. Agent 1, or Agent 2, etc.) that the participant had selected earlier appeared next to the typing area and gave step by step instructions on how to use that particular Ms Office Word function to complete the given task.

SMOS also included features that could allow a participant to manipulate the agent's behaviours. Referring back to Fig. 6.6, note that there were three buttons situated above the agent, namely Play, Pause and Stop. The Play button allowed a participant to replay the agent's instructions in case he/she did not understand. This button could also be used to restart the agent after it was paused. A participant could use the Pause button to stop the agent for a moment or for lengthier periods of time. With the help of the Stop button, a participant could stop the agent from giving instructions. Likewise, below the agent there were also three buttons (i.e. Seek to, Speed, and Volume). These three buttons were slider controls. The Seek to button enabled the participant to locate (search for) a specific moment in time in the agent's instructions. Participants could use the Speed button to control the speed (slower or faster) of the agent's instructions. The Volume button could be used to control the volume (increase or decrease) of the agent's voice.

Changing from one kind of agent to another. After a participant had completed the given tasks with one agent, he/she needed to do the following in order to select another agent:

- After first ensuring that changes were saved, the Ms Office Word Simulation Window needed to be closed.
- The participant had to return to the 'Choose an agent' screen to select the new agent.
- A double click on the Ms Office Word Simulation shortcut opened the Ms Office Word simulation window again.

Fig. 6.7 A participant
interacting with a
pedagogical interface agent

Figure 6.7 shows a participant using SMOS during the training. It illustrates the interaction between a participant and one of the pedagogical interface agents incorporated in SMOS.

The next section will provide a detailed discussion of the different usability techniques used by the researchers to carry out the usability testing in the adult learning environment.

6.4 Research Design and Methodology

6.4.1 Research Design

The researchers chose a mixed methods triangulation design (a kind of design that uses concurrent data collection techniques for collecting quantitative and qualitative data) that combined qualitative and quantitative analysis as best appropriate research design.

As a result, quantitative analysis was used to:

- Obtain data about the efficiency (competency) and effectiveness (capability) of the participants in performing different tasks with pedagogical interface agents.
- Collect data on the relationship between diverse variables used in the study.
- Present data in a form of statistics and aggregated data.

In addition, for the purpose of this study, qualitative analysis was used to:

- Describe the satisfaction of the participants with pedagogical interface agents.
- Obtain subjective impressions or preferences among the participants about their interaction with a variety of pedagogical interface agents.
- Build theories.

Section 6.4.2 introduces and discusses the research methodology used for collecting data needed for this research study.

6.4.2 Research Methodology

Research methodology in this study refers to the different techniques that were utilized for collecting both qualitative and quantitative data.

Data Collection methods. Jonson and Christensen [32] identified the six most common methods of data collection, namely usability tests, questionnaires, interviews, focus groups, observation, and secondary or existing data. The data collection methods used in this study involved usability tests, questionnaires, observation and interviews. Usability tests were conducted in order to evaluate the extent to which a variety of pedagogical interface agents could assist participants in acquiring basic computer skills.

Self-designed questionnaires were utilized to collect relevant opinion-related data from all the participants. The questionnaires included both structured and unstructured types of questions. The question format consisted of a mixture of both the Likert scaling and open-ended questions, and the questionnaires were administered in English.

All participants completed two types of questionnaires, namely a pre-training and a post-test questionnaire. Each of these questionnaire types served to collect different kinds of data for different purposes in this study. The next two subsections will provide a more detailed discussion on these two types of questionnaires.

Pre-training questionnaires. The pre-training questionnaires referred to questionnaires that were completed by all participants prior to the basic computer literacy training sessions. They consisted of 30 questions which were divided into four parts, namely Personal Information, Computer Experience, Computer Characters, and General Issues. The following were the main reasons for requesting the participants to complete the pre-training questionnaire:

- To gather personal information about participants (i.e. age, gender, qualification, etc.).
- To gather participants' opinions on computers and also evaluate their computer experience.
- To gain insight with regard to the participants' knowledge, views and attitudes towards pedagogical interface agents.
- To gather comments from participants on their participation in the study.

Questions contained in pre-training questionnaires allowed the researchers to do training needs analysis in order to determine the participants' existing skills, knowledge and abilities.

Post-test questionnaires. The post-test questionnaires referred to questionnaires that were completed by all participants after completing the usability test. The researchers designed two kinds of post-test questionnaires: one for the test group participants who were exposed to agents, and the other for the control group who was not exposed to agents.

The post-tests questionnaires for the test groups consisted of 43 questions, while those for the control group participants consisted of 31 questions. Questions included in both post-test questionnaires focused on measuring the participants' personal experiences, including their satisfaction levels, regarding the tools that they used during the usability test sessions. For the test groups participants, the term 'tools' is used to refer to a variety of pedagogical interface agents that they used during the usability testing sessions. With regard to the control group participants, 'tool' referred to Ms Office Word.

Observations and Interviews. Besides the use of the above-mentioned questionnaires, the researchers also used observation and interviews as additional data collection methods to supplement questionnaires. Participants' body language and physical use of the system were observed, while problems, irregularities and inconsistencies were clarified during interviews.

Pilot Study. Data plays a very significant role in a study and it was inevitable to test the appropriateness of the chosen data collection methods. Therefore a pilot study was undertaken before the full scale study. From the researchers' point of view the purposes of the pilot study were the following:

- To identify the different agents' shortcomings.
- To test if employed data collection techniques were able to help achieve the objectives of the study as planned (refer to Sect. 6.2.2).
- To check if the chosen data collection methods best fitted the problem as well as being consistent and suitable to answer the investigated research question.
- To check if the questionnaires were clear and answerable.

Notes taken by the researchers while observing participants during the pilot test, including verbal and written feedback received from participants, revealed a number of problems with regard to some agents and parts of the questionnaires. Thus, the researchers' notes and participant feedback were used to refine the post-test questionnaires, and also to improve some of the pedagogical interface agents' functionalities.

The identification and recruitment of potential participants representative of the population is discussed in the next section.

Population and Sampling Procedure. In this study, the target population consisted of adult learners. Therefore, it was imperative to first of all define the characteristics of adult learners that needed to form part of the population. The following were the list of characteristics used to recruit the adult learner participants:

- Goal-oriented
- Relevancy-oriented
- Little or no formal post-school educational level
- Able to read and write in English
- Little or no previous experience of computers.

The researchers identified the Mangaung University of the Free State Community Capacity Programme (MUCCP) as an ideal place to find the target population. MUCCP is located in the Pelindhaba Township of the Mangaung Local Municipality in Bloemfontein, which is the provincial capital of the Free State

Province located in the centre of SA. MUCCP is a partnership between communities, higher education institutions and the services sectors formed in 1991 [33]. The main objective of this partnership was to promote sustainable livelihood for previously disadvantaged and unemployed adults in Mangaung, particularly around the townships. In view of the fact that MUCCP was selected as the best place to conduct the main study, the researchers met with the MUCCP manager to explain the purpose of the study, seek her approval to conduct the study, request her assistance in the recruitment process of the participants that would form part of the main study, and to request the use of a venue for conducting the training.

A sample refers to a number of individuals selected from a specific population group to represent a large group from the population from which it was drawn [34]. In the context of this study, a sample is a small portion representative of the target adult learner population. The sample consisted of 103 adult learners from MUCCP. All 103 participants were requested to fill in a consent form and a pre-training questionnaire. By completing the consent form, participants were informed about all the potential risks and costs involved in the study [35]. It also guaranteed their privacy and right to anonymity. It explained in detail what their participation would entail.

Data collected from the pre-training questionnaires helped to decide which topics needed to be covered during the training.

Basic Training Sessions. The basic training referred to the initial training that all participants had to undergo. All of the 103 participants had similar characteristics (e.g. little/no previous knowledge of computers). Before the initial training started, participants were randomly assigned to 13 groups. These 13 groups were further divided: 4 groups were referred to as the *control groups*, while the other 9 groups were considered the *test groups*. The control groups were the groups of participants that received normal Ms Office Word training. The test groups referred to the groups of participants who were exposed to the agents as well during their training sessions.

The aim of the basic training was to introduce participants to the computer environment and to teach them useful computer skills and basic Ms Office Word skills. The reason for doing so was to expose all participants to the computer atmosphere and to familiarize them with the Ms Office Word working environment. The idea was also to level the playing field in terms of computer knowledge.

The basic training sessions were conducted in the Computer Laboratory located on the premises of the MUCCP. The computer laboratory contained 10 personal computers (PCs) allowing groups of maximum 10 participants only as each participant was allocated his/her own PC for the duration of a training session. Due to the limited number of PCs available, participants were notified of the dates and the starting times of their respective training sessions by means of SMSs. Each participant was requested to confirm his/her attendance.

The training material was compiled by the researchers based on participants' computer experience as collected in the pre-training questionnaires. As a result, the contents of the basic training material did not include all the basic Ms Office Word tasks, but rather consisted of a selected number of tasks that the researchers considered to be relevant for the scope of this study. The basic training material included both theory and hands-on practical work. One group per week was trained. The basic training per group took 4 days (the 5th day (Friday), was reserved for the usability testing session). It took 13 weeks in total to carry out the basic training for all 13 groups of participants. Days 1 and 2 consisted of teaching and hands-on exercises performing Ms Office Word tasks that were demonstrated by the researchers. Participants were advised to practice what had been demonstrated under the supervision of the researchers. The researchers' demonstrations primarily involved an introduction to the Windows environment, keyboard and mouse skills, and lastly focused on the Ms Office Word environment. Table 6.1 shows a few examples representing Windows and Ms Office Word tasks included in the basic training material.

In Table 6.1 'tasks' refer to what is supposed to be done, whereas the term 'instructions' is used to refer to how a particular task should be done.

Following the demonstration, on days 3 and 4 hands-on laboratory exercises and training tasks were given to participants as individual work to be done under the supervision of the researchers. These laboratory exercises and tasks were related to the training material presented during days 1 and 2. While doing their various exercises and training tasks participants were allowed to ask questions or identify areas where further explanations were needed.

Test group versus control group training. The same basic training material, hands-on laboratory exercises and training tasks were used to train both test and control groups. The most important differences between these two groups were in the manner in which their basic training was conducted, as well as in the ways in which they performed their hands-on laboratory exercises and training tasks. The test group participants were trained under two conditions: with or without agents.

Table 6.1 Basic training tasks

Tasks	Instructions
Clicking	Pressing the left mouse button down once and releasing it quickly
Double clicking	Clicking the left mouse button twice in quick succession
Capitalizing a letter	Hold down the Shift key while you press the key for that letter
Start new paragraph	Press the Enter key twice
Insert new text	Move the cursor at the specific location (the insertion point) where you would like to insert the new text on the text area and click
Bold	Select the text you want to appear bold by highlighting it. To make the text that you had selected bold, click the Home tab on the Ribbon, go to Font group, and click the bold button
Exit word	Click the Microsoft Office button. A menu appears. Click Exit Word, which you can find in the bottom-right corner
Open a saved file	Click the Microsoft Office button. A menu appears. Click Open. The Open dialog box appears. Use the Look In field to move to the folder in which you saved the file. Click on the file. Click Open
Delete entire word or multiple words	Select the entire word or multiple words to be deleted by highlighting them. Press the Delete key on the keyboard

With agents meant that the group participants were introduced to agents and worked with the different kinds of pedagogical interface agents during their training. *Without agents* meant that the group participants were not introduced to agents during their training. The reason for introducing the test group to agents was to equip them with the necessary skills that would enable them to make use of the agents when performing various training tasks. The control group participants were trained under only one condition, in other words, they only received basic Ms Office Word training. They were not exposed to the agents and did not know anything about the agents. The reason for dividing participants into two groups was to use the control group as benchmark to measure how participants in the test groups performed. This allowed for determining the effect of using a variety of pedagogical interface agents during training.

The different approaches used in training also influenced the way participants of these two groups performed their hands-on laboratory exercises and training tasks. The test group participants did some of their laboratory exercises and training tasks with the assistance of the agents and other tasks without the agents. On the other hand, the control group participants performed all their laboratory exercises and training tasks without the agents since they did not have any knowledge about them.

The next section will explain the usability testing.

Usability Testing. The focus of the research in terms of the usability testing was on efficiency, effectiveness, and satisfaction. This involved measuring the performance of participants while undertaking a predefined set of Ms Office Word tasks during the usability test.

Two systems were used to conduct the usability testing, namely a normal and a simulated Ms Office Word environment (SMOS) as discussed in Sect. 6.3.3.

Experimental Design. The basic training equipped participants with the necessary skills for the usability test experiment. The usability testing was done immediately after the basic training. The Pre-test Post-test control group experimental design appeared to be the best suitable approach to conduct this usability test experiment. This kind of experimental design usually involves randomly assigning participants into test groups and control groups. Once the groups have been established all the participants from both groups are pre-tested; then the experimental treatment condition is administered to one group (i.e. test group). Afterwards the post-test assessments of both groups are performed.

In this research study, participants were also randomly assigned to either a test group or a control group. All the participants received Ms Office Word training (pre-test) while only the test groups were introduced to pedagogical interface agents (experimental treatment). Afterwards both groups were given tasks to perform (post-test).

For the purpose of the usability test experiment, the researchers designed 11 simple tasks to represent an opportunity for participants in both control and test groups to use their respective systems. Participants in both groups were instructed to complete the 11 tasks, either working with the assistance of pedagogical interface agents for the test groups, or using Ms Office Word's normal features for participants in the control groups. The test groups performed 10 tasks using ten different kinds of pedagogical interface agents and 1 task (i.e. task 11) without the assistance of an agent.

The experiment was a mixture of *between-group* and *within-group* designs. It was firstly a 'two-groups between-group' design. The two groups consisted of the test group and control group. The test group was exposed to pedagogical interface agents and control group was not exposed to the agents. The between-group independent variables were the tasks that were performed either with pedagogical interface agents or without and the dependent variable was usability performance. The researchers manipulated the between-group independent variable in order to compare the usability performance of these two groups (test groups and control group).

Additionally, the experiment was a 'two-groups within-group' design. The within-group independent variable was the tasks and the within-group dependent variable was usability performance. The researchers manipulated the within-group independent variable in order to compare the usability performance of participants within each particular group (control group participants and test group participants).

The following standards of measurement metrics were used in this study:

- Task completion success
- Number of participant errors
- Participant satisfaction.

These three standards of measurement were considered as measurable performance indicators. They were used in the current study for measuring the effectiveness (being able to successfully complete a task) and efficiency (the effort required to complete a task) of a variety of pedagogical interface agents [36]. Additional efficiency metrics exist, namely time taken to complete a task, as well as the number of actions or steps taken by a participant to perform a task. However, these two metrics were not included in this study. The number of errors was the only efficiency metric that was taken into consideration. Measuring participants' satisfaction was done via qualitative measures.

Conducting the Test Experiment. As mentioned before, the test group had a total of 72 participants and the control group was made up of 31 participants. The researchers designed different task sheets for each group. They contained the same 11 tasks but different instructions on how to perform them were given.

Each participant was handed a task sheet and was instructed to follow the instructions on it when performing the tasks. Participants received no assistance from the researchers in performing the tasks. The researchers were observing participant behaviour and also took notes. If they needed assistance, test group participants were instructed to make use of the agents' assistance and the control group participants to use normal Ms Office Word assistance.

After completing their usability test tasks, all the participants were asked to complete post-test questionnaires. Different questionnaires were designed for the two groups as they performed their tasks in different ways. Usability metrics and the two post-test questionnaires completed by participants allowed the researchers to collect both quantitative and qualitative data about the participants.

The analysis of the data collected, including the interpretation thereof, will be discussed in Sect. 6.5.

6.5 Data Analysis and Interpretation

Section 6.4 discussed the research design used to carry out the current study. The utilised data collection strategies were also outlined. This section will present, analyse, interpret and discuss the collected data. The main focus of the analysis will be on the data obtained through the usability tests, questionnaires and observations. Lastly, a comparison of the data obtained will be presented as well.

6.5.1 Demographic Data

As previously mentioned the pre-training exercise was used to collect general information concerning the study participants. This information included their biographic (personal) information and computer experience. Each of these will be explained briefly.

Biographic information. Bearing in mind that this study was geared towards a population of adult computer illiterate users with little or no post-school training, the personal information captured included gender, age, home language, education level and occupation.

Gender, Age, and Home Language. Of the 103 participants 62 % were female and 38 % were male. With regard to their ages, 14 % were younger than 20, 53 % were between 20 and 30 years of age, 25 % were between the ages of 31 and 40, and 4 % were older than 40. Four percent did not indicate their age.

The majority of the participants (48 %) indicated Sesotho as their home language, 27 % Setswana, 17 % Xhosa and 1 % Zulu. The rest did not specify their home language.

Education Level and Occupation. The majority of participants (68 %) was in possession of a matric certificate as highest qualification, 22 % passed standard 9 (Grade 11), and 6 % passed standard 8 (Grade 10). Additionally, 2 % of participants passed Grade 9, 1 % was N3 engineering certificate holders, and the other 1 % did not specify their educational levels.

A total of 20 % indicated that they were unemployed. The employed participants held a variety of occupations. These consisted of 15 % volunteers, 4 % cleaners, 3 % self-employed, 3 % students, 2 % care workers, 1 % skills coaches, 1 % FM presenters, 1 % cashiers, 1 % stock checkers, 1 % blenders, and 1 % food assistants.

Useful data concerning participants' computer experience was collected also.

Computer Experience. Computer experience questions related to the ability and time spent using computers, the purpose of using computers and the ability to start up a computer.

Participants were asked to indicate if they had ever used a computer before. 53 % of them had done so, whereas the remaining 47 % had not. Among the participants who had used a computer before, 18 % used it on a monthly, 15 % weekly, 14 % daily, and 2 % on an occasional basis. The other 51 % did not specify/did not respond to the question.

Participants who have used a computer before were also asked to indicate the purpose of their computer usage. 33 % indicated that the computer was used for typing a document, 22 % for playing music, 21 % for playing games, and 12 % for browsing the Internet. The other 9 % of responses indicated that the computer was used for sending email, 2 % did not specify, and 1 % used it for Pastel. Most participants (57 %) indicated that they knew how to switch on a computer, while 18 % admitted that they did not.

The Sect. 6.5.2 presents the usability performance data analysis.

6.5.2 Statistical Analyses of Usability Performance Data

The usability metrics captured while participants were working on the tasks assigned to them included efficiency (number of errors) and effectiveness (task success). These were useful in determining whether there were usability performance differences between participants in the test and control groups or not. Of the 31 participants in the control groups and 72 in the test groups, 4 in the test groups failed to submit the corrected documents and were not included in these results. In other words, with regard to the usability performance data only 68 test group participants were included.

The efficiency (number of errors) statistical data analysis will be discussed first.

Efficiency. Number of errors is a count. During the usability test assessment it referred to the number of missed steps when performing a given task. The number of errors was measured using a binary technique whereby a 0 was allocated to failure and a 1 to success. The participants received a 0 for each 'no error' step and a 1 for each incomplete step or for a step not attempted. Two kinds of analysis were performed, namely the total number of errors and number of errors per each individual task.

The total number of errors. An independent-samples t-test was conducted to compare the total number of errors that participants from both groups, the test group (i.e. participants who worked with agents) and the control group (i.e. participants who did not work with agents), made when performing the usability assessment. The results of the t-test (where $\alpha = 0.05$ and the P-value 0.001) indicated that there was a significant difference in the total number of errors made in the usability test assessment by the test group participants compared to those in the control group. In other words, the results revealed that the participants who worked without agents' assistance made significantly more errors in the assessment than those who used the agents.

All participants performed eleven tasks during the usability test assessment. The number of errors for each of these eleven tasks was analysed also.

The number of errors per individual task. Independent-samples t-tests were also conducted to compare the scores of individual tasks in the usability test assessment for those participants who used the agents and those who did not.

Considering the P-values ($\alpha = 0.05$) for the 11 tasks displayed in Table 6.2, only four tasks (3, 5, 6, and 11) among them showed that there was significant statistical differences in errors (scores) for those who used the agents and those who did not. For each of these named tasks the control group participants made significantly more errors in the usability test assessment than the test group participants.

It is important to note that the test group participants did not use an agent in task 11, but they still performed better than the control group participants. This is a positive indication for the test group participants and for the usefulness of agents in this situation.

Additionally to the efficiency statistical data analysis, the participants' task success was analysed statistically to determine any statistical significance in the effectiveness.

Effectiveness. Effectiveness refers to being able to successfully complete a task or a step in a given task. In the context of this study, success meant that a participant managed to successfully complete at least one step when performing a task. The task success was measured using a binary approach (1 for success or 0 for failure). The participants received a 1 for each step completed successfully in a given task and a 0 for an incomplete step or a step not attempted.

A Chi square test was conducted in order to compare the proportion of test group and control group participants who had at least something right/nothing right in a task. The Chi square analysis results are displayed in Table 6.3.

Note that in Table 6.3, there are no statistical analysis results for tasks 2, 3, 4, 5, 7, 8, 10 and 11; for each of these the 'minimum expected cell frequency' was less than five, therefore these could not be analysed using a Chi square test.

Similar to the t-test, the Chi square test also indicates a statistical significant result if the P-value for the test is less than 0.05. Among the three tasks (1, 6, and 9) shown in Table 6.3, task 6 was the only one where the Chi square test for independence (with Yates Continuity Correction) indicated a significant difference between the proportion of participants in the test group and control group who had at least something right/nothing right in task 6. This implies that a higher proportion of test group participants had at least something right when performing task 6, when compared with control group participants.

Bearing in mind that the main purpose of this study was first of all to determine whether pedagogical interface agents could enhance the computer literacy training, and subsequently the employment possibilities, of adult computer illiterate users, the next step would be to investigate what these agents should look like and how

Table 6.2 Comparison of the number of errors per individual task

Task	Descriptive statistics	Comparison	
		With agents (n = 68)	Without agents (n = 31)
1. Open the document, use spelling or grammar check, save, and exit	Mean	0.18	0.16
	SD	0.384	0.374
	P-value	0.854	
	Statistical significance	No significance	
2. Open the document, make text bold and italicized, align text, save, and exit	Mean	0.59	0.84
	SD	0.777	1.157
	P-value	0.208	
	Statistical significance	No significance	
3. Open the document, change text font, font colour, save, and exit	Mean	0.12	0.61
	SD	0.406	0.715
	P-value	0.001	
	Statistical significance	Significance	
4. Open the document, underline text, insert the word, save, and exit	Mean	0.24	0.58
	SD	0.601	1.057
	P-value	0.098	
	Statistical significance	No significance	
5. Open the document, insert blank lines, insert picture, align centre, save, and exit	Mean	0.35	1.19
	SD	0.617	0.833
	P-value	0.000	
	Statistical significance	Significance	
6. Open the document, delete text, undo, redo, save, and exit	Mean	0.28	1.26
	SD	0.861	1.437
	P-value	0.001	
	Statistical significance	Significance	
7. Open the document, cut, move text, save, and exit	Mean	0.54	0.97
	SD	0.679	1.110
	P-value	0.056	
	Statistical significance	No significance	
8. Open the document, make bulled list, italicize, save, and exit	Mean	0.18	0.26
	SD	0.455	0.575
	P-value	0.449	
	Statistical significance	No significance	
9. Open the document, align centre, underline text, save, and exit	Mean	0.18	0.45
	SD	0.455	0.810
	P-value	0.085	
	Statistical significance	No significance	

(continued)

Table 6.2 (continued)

Task	Descriptive statistics	Comparison	
		With agents (n = 68)	Without agents (n = 31)
10. Open the document, search and replace text, save, and exit	Mean	0.04	0.16
	SD	0.207	0.374
	P-value	0.110	
	Statistical significance	No significance	
11. Open the document, select block of text, change the font size, Save and exit	Mean	0.28	0.65
	SD	0.619	0.915
	P-value	0.049	
	Statistical significance	Significance	

Table 6.3 Comparison of effectiveness among tasks

Task	Selective information	Total participants in both groups = 99
1. Open the document, use spelling or grammar check, save, and exit	Chi square (Continuity Correctionb)	0.000
	P-value	1.000
	Statistical significance	No significance
6. Open the document, delete text, undo, redo, save, and exit	Chi square (Continuity Correctionb)	8.849
	P-value	0.003
	Statistical significance	Significance
9. Open the document, align centre, underline text, save, and exit	Chi square (Continuity Correctionb)	1.096
	P-value	0.295
	Statistical significance	No significance

they should interact with the user. In an attempt to answer these questions and to elicit participants' opinions regarding their experiences during the usability test assessments, participants' satisfaction levels were measured by means of post-test Likert scale questionnaires. The questionnaires consisted of 5 options: strongly disagree, disagree, not sure, agree, and strongly agree, and will be discussed in the following section.

6.5.3 Post-test Questionnaire Analysis

Two types of post-test questionnaires were used. One was specifically aimed at the test group participants who were exposed to agents (refer to Appendix A), and the other kind towards the control group, who was not (refer to Appendix B). Unlike the performance data analysis, the four participants in the test group who failed to submit the correct documents were also counted and included in the satisfaction analysis as they worked with agents and could, therefore, give feedback regarding their personal experiences with the agents.

Control group post-test questionnaire. The control group participants used a normal Ms Office Word environment to perform their usability tests. The responses for the two questionnaire categories namely agree and strongly agree, were combined for the purpose of this discussion.

In analysing the satisfaction levels of the control group participants (and subsequently the test group as well), n was used to represent the total number of participants who responded to that particular statement. M represented the Mean (average). Min and Max represented minimum and maximum respectively, whereas SD represented the standard deviation.

As shown in Table 6.4, the majority of participants (97 %) in the control group by far strongly agreed/agreed that they were able to use Ms Office Word successfully. Furthermore, they also felt that their experience with Ms Office Word encouraged them to learn about other computer programs, as well as to encourage their friends to learn new concepts. Most participants (93 %) in the control group strongly agreed/agreed that it was exciting working with Ms Office Word; 90 % that working with Ms Office Word made them change their attitude towards

Table 6.4 Summary of control group participants' post-test questionnaire responses

Statement	Strongly agreed/agreed (%) n = 31	Mean (M)	Minimum (Min)	Maximum (Max)	Standard deviation (SD)
I was able to use MS office successfully	97	4.39	2	5	0.667
It was exciting working with MS office	93	4.50	2	5	0.745
Working with MS office made me change my attitude towards computers	90	4.32	1	5	0.945
Interaction with MS office was easy	87	4.239	2	5	0.762
I really had to concentrate working with MS office	87	4.16	1	5	0.932
I felt frustrated while working with MS office	24	2.45	1	5	1.352
I felt nervous when working with MS office	20	2.13	1	4	1.137

computers, 87 % that the interaction with Ms Office Word was easy, but that they really had to concentrate to do the tasks. Few participants (24 %) strongly agreed/ agreed that they felt frustrated while working with Ms Office Word, while 20 % strongly agreed/agreed that they felt nervous.

To investigate the influence of exposure to agents, the test group post-test questionnaire will now be discussed.

Test group post-test questionnaire. The test group participants used a variety of pedagogical interface agents to assist them in performing their usability tests.

The post-test results summarised in Table 6.5 reveal that the vast majority of the test group participants (97 %) strongly agreed/agreed that they were able to use agents successfully and that they trusted the advice from agents. 96 % strongly agreed/agreed that the agents' hints helped them to feel more confident about their computer skills, and a total of 94 % that it was exciting working with the agents and that they also could encourage friends to use agents when learning new concepts. 88 % of test group participants strongly agreed/agreed with the following statements:

- Agents made them change their attitude towards computers
- Their experience with agents encouraged them to find out more about them
- They would consider using agents when learning concepts in real life.

As far as their concentration was concerned, 84 % strongly agreed/agreed that they really had to concentrate to work with the agents, and 80 % found the agents to be user friendly. Only 21 % of participants strongly agreed/agreed that they felt nervous working with the agents, and the other 14 % that they felt frustrated.

Individual preferences regarding agents. Participants were asked questions about their personal preference regarding the pedagogical interface agents. About 51 % of participants indicated that they preferred a cartoon agent to a realistic agent. Some 49 % indicated that they preferred a male agent to a female agent. The other 46 % chose a text agent to a 'text and audio' agent. 42 % of the participants stated that they preferred a dog agent to a human agent.

Referring to Table 6.6 the following was noted:

- The female text agent was preferred in terms of appearance (20 % of participants selected this option). This was followed by the male cartoon dog agent (16 %). The agent least liked in terms of appearance was the male realistic dog agent.
- The most popular agent in terms of voice was the female 'text and audio' agent (18 %), followed by the male cartoon dog agent (16 %). The two least liked agents were the male realistic dog and the female realistic dog agents (both 3 %).
- The preferred agent based on movement was the female cartoon human agent (23 %). Here the second place went to the male cartoon dog agent (16 %), followed by the male realistic dog agent (14 %). In terms of movement the least liked agents were the male text agent, the female text agent, and the female realistic dog agent who scored 5 % each.

Table 6.5 Summary of test group participants' post-test questionnaire responses

Statement	Strongly agreed/agreed (%) n = 72	Mean (M)	Minimum (Min)	Maximum (Max)	Standard deviation (SD)
I was able to use the agents successfully	97	4.18	2	5	0.699
The agents' hints helped me to feel more confident about my computer skills	96	4.55	3	5	0.580
It was exciting to work with agents and I will encourage my friends to learn new concepts	94	4.40	2	5	0.689
Agents made me change my attitude towards computers, encouraged me to find out more about them and I will consider using agents again when learning concepts in real life	88	4.37	2	5	0.722
I really had to concentrate working with the agents	84	4.10	1	5	0.923
I found the agents to be friendly	80	4.17	1	5	1.084
I felt nervous when working with the agents	21	2.22	1	5	1.259
I felt frustrated working with the agents	14	2.10	1	5	1.241

Table 6.6 Preference levels in terms of agents

Agents	Preference criteria (Percentage of participants)		
	Appearance (%)	Voice (%)	Movement (%)
Female text agent (Agent 1)	20	11	5
Male text agent (Agent 2)	9	15	5
Female text and audio (Agent 3)	11	18	7
Male text and audio (Agent 4)	9	6	7
Female cartoon human (Agent 5)	9	8	23
Male cartoon human (Agent 6)	9	11	9
Female cartoon dog (Agent 7)	7	8	11
Male cartoon dog (Agent 8)	16	16	16
Female realistic dog (Agent 9)	7	3	5
Male realistic dog (Agent 10)	4	3	14

The majority of participants enjoyed working with all the agents, where the male realistic dog scored the highest (95 %) and the male text agent the lowest (76 %).

Participants were also asked to provide their individual overall preference with regard to the pedagogical interface agents they used. Although the female text agent scored the highest in terms of appearance, the female 'text and audio' agent in terms of voice, and the female cartoon human agent in terms of movement (see previous section), the overall favourite was the male cartoon dog. For the reader's convenience the preference ratings are summarised in Table 6.7.

Comparison between the test and the control groups. The participants from the test and the control groups were also compared in terms of the following: (1) how much they enjoyed being part of the study, (2) how much they learnt from the study overall, and (3) if they would like to participate in a similar study in future.

All the participants in both groups indicated that they enjoyed the study.

An independent-samples t-test was conducted to compare the mean scores on "amount learnt from the study overall" for the test and control group participants. The results of the t-test are shown in Table 6.8.

Table 6.7 Agent preference ratings in terms of agent characteristics

Agent	Overall	Appearance	Voice	Movement
Male cartoon dog (Agent 8)	1	2	2	2
Female text (Agent 1)	2	1	4	8
Female text and audio (Agent 3)	3	3	1	6
Female cartoon human (Agent 5)	4	4	6	1
Male cartoon human (Agent 6)	5	4	4	5
Male text and audio (Agent 4)	6	4	8	6
Female cartoon dog (Agent 7)	7	8	6	4
Male text (Agent 2)	8	4	3	8
Male realistic dog (Agent 10)	9	10	10	3
Female realistic dog (Agent 9)	10	8	9	8

Table 6.8 Comparative amount learnt from the study overall

Selective information	Comparison of amount learned from the study overall	
	With agents (n = 72)	Without agents (n = 31)
Mean	1.26	1.20
SD	0.640	0.484
P-value	0.662	
Statistical significance	No significance	

The P-value (with $\alpha = 0.05$) in Table 6.8 indicates that there is no statistical significant difference between the two groups in terms of amount learnt from the study overall. In terms of their own perception of learning participants in the test group did not learn more from the study than the control group participants.

Both groups indicated that they would like to participate in a similar study in future.

6.6 Conclusions

This section summarizes the findings and the conclusions reached in this research study.

6.6.1 Overview

Technological illiteracy and lack of formal education among the adult population in developing third world countries have been a major challenge for most of the third world governments. Many jobs require the use of computers as part of everyday tasks. There is a need to find innovative approaches to ease technology transfer to adults with little or no technological background or formal post-school education in third world countries. This can assist them in becoming part of the workforce, raising their self-confidence and feelings of worthiness, and thus making them an integral part of society.

For several years research has been carried out regarding the potential of pedagogical interface agents to promote learning. One common finding among these studies was that pedagogical interface agents can improve a student's learning, engagement and motivation. Additionally, the majority of research studies have focused on childhood to undergraduate, college-aged populations. Little is known about the benefits of pedagogical interface agents in adult learning environments—hence the incorporation of pedagogical interface agents into adult learning environments continues to be an open question.

6.6.2 Findings

The aim of this research study was to determine the usability effect of pedagogical interface agents on adult computer literacy training. The sample population was divided into two groups where one was exposed to agents during their training (test group) and the other (control group) was not. Both groups were given 11 basic Ms Office Word tasks to complete during a usability test. The test group participants did 10 tasks using ten different kinds of agents and one task without agent assistance. The control group participants did all eleven tasks without agents' assistance. However, they were allowed to use the original Ms Office Word Help menu. The aim of capturing usability metrics was to compare the usability performance (in terms of efficiency and effectiveness), as well as the user satisfaction of using pedagogical interface agents during adult computer literacy training, between the two mentioned groups of participants.

Usability performance. Two hypotheses were formulated in order to determine the usability performance. These two hypotheses were related to efficiency and effectiveness.

Efficiency. $H_{0,1}$: There is no difference in the efficiency when using a pedagogical interface agent to train adult computer illiterates when compared to using traditional educational training techniques.

The statistical analysis for the two groups (i.e. test and control) relating to efficiency was compared in terms of the 11 tasks. The following outcome variables were analysed:

- The total number of errors
- The number of errors per each individual task.

The usability performance data was collected using a mixed methods triangulation design that combined qualitative and quantitative methods. An independent-sample t-test was conducted to compare the total number of errors made in the usability assessment by participants in the test and the control groups. Likewise another independent-sample t-test was conducted to compare the scores of the number of errors in each individual task in the usability assessment by participants who used the agents and those who did not.

As far as the total number of errors was concerned: there was a statistical significance in the total number of errors made, therefore $H_{0,1}$ was rejected.

In terms of the number of errors per individual task there was a statistical significance for four tasks, namely tasks 3, 5, 6, and 11. Therefore, $H_{0,1}$ was rejected for the named tasks. As there was no statistical significance in the number of errors for the rest of the tasks, $H_{0,1}$ was not rejected.

With regard to efficiency, it can be concluded that the use of the pedagogical interface agents helped participants to reduce the number of errors (i.e. number of errors in total and per individual task) while performing various tasks. Additionally it should be noted that the level of support that agents provided to the participants varied from one agent to another.

Effectiveness. $H_{0,2}$: There is no difference in the effectiveness when using a pedagogical interface agent to train adult computer illiterates when compared to using traditional educational training techniques.

The effectiveness statistical analysis for the two groups (i.e. test and control) was compared in terms of the 11 tasks. Task success as outcome variable was analysed by means of a Chi square test to compare the proportion of test and control group participants who had at least something right/nothing right in the task.

As far as task success was concerned, task 6 was the only task that indicated a statistical significance, therefore $H_{0,2}$ was rejected for this task. While there was no statistical significance in the task success for tasks 1 and 9, $H_{0,2}$ was, therefore, not rejected for these two tasks.

A propos of effectiveness, it can be concluded that pedagogical interface agents' assistance enhanced the capability of participants to successfully perform various tasks. Furthermore, it should be pointed out that the degree of competence that participants showed while working with the various pedagogical interface agents might have differed from one agent to another.

The user satisfaction data is discussed in the next section.

User Satisfaction. $H_{0,3}$: There is no difference in the user satisfaction when using pedagogical interface agents to train adult computer illiterates when compared to using traditional educational training techniques.

The statistical analysis for the user satisfaction was compared between the test and the control groups with regard to the 11 tasks. The amount learnt from the overall study was analysed as outcome variable by means of an independent-sample t-test.

There was no statistical significance in the amount learned from the study overall for the test group and control group participants. Therefore $H_{0,3}$ was not rejected.

In terms of user satisfaction, from the participants' perceptions it can be concluded that the use of the pedagogical interface agents did not influence their learning gains in terms of the amount learned from the study overall. It should also be noted that the participants' views about or understanding of the outcome variable might have differed from one group of participants to another.

In relation to the changes in participants' knowledge, attitude and aspiration based on their computer literacy training with pedagogical interface agents, participants strongly agreed/agreed that

- Agents made them change their attitude towards computers;
- Their experience with agents encouraged them to find out more about them;
- They would consider using agents when learning real life concepts.

This indicates that the use of pedagogical interface agents positively influenced the changes in participants' knowledge, attitude, and aspiration.

6.6.3 Possible Future Research

The findings of this research study have shown that the incorporation of educational interface agents is a promising step towards suggesting ways to improve adult computer literacy in developing countries.

Future extensions to this research project might be to explore the possibility of:

- Utilizing other participant demographics such as young illiterate adults;
- Participants working on complex tasks;
- Using a greater variety of educational interface agents;
- Including more efficiency measurement metric attributes such as the time taken to complete a task, and the number of actions or steps taken by participants to perform a task.

Acknowledgments The authors would like to thank the Telkom Centre of Excellence at the Department of Computer Science and Informatics, University of the Free State, for partially funding this research. Thanks to Mr. Casper Wessels from the Department of Information Technology, Central University of Technology, Free State for providing computers that were used to carry out this research. Our thanks also go to the adult learners from MUCCP for their willingness to participate in this research. Mrs. Suezette Opperman is also thanked for the language editing, and Dr. Melody Mentz for assisting with data analysis.

Appendix A

Post-test Questionnaire (without agents)
The purpose of this questionnaire is to elicit your personal opinions of the Microsoft Office environment you had worked with while carrying out different tasks for evaluation purposes. Please answer all the questions.

For question 1 to question 23 indicate rate your opinion on a scale of 1–5, place a circle around the appropriate number, where: 1 = "Strongly disagree", 2 = "Disagree", 3 = "Not sure", 4 = "Agree", 5 = "Strongly agree"

No.		SD	D	NS	A	SA
Learning contents provided by the microsoft office						
1.	Microsoft office had functionalities I expected it to have	1	2	3	4	5
2.	Microsoft office environment was easy to use	1	2	3	4	5
Effectiveness of the microsoft office help function						
3.	I found the microsoft office help function to be useful	1	2	3	4	5
4.	Microsoft office help function provided me with all the necessary information	1	2	3	4	5
5.	Microsoft office help function helped me to quickly learn how to perform a particular task	1	2	3	4	5
6.	Microsoft office help function helped me to recall the different steps involved for a particular task	1	2	3	4	5
7.	Microsoft office help function helped me to complete my tasks quicker	1	2	3	4	5
8.	Microsoft office help function was very practical	1	2	3	4	5
9.	Microsoft office help function assisted me to identify my mistakes when performing a task	1	2	3	4	5
10.	With microsoft office help function, it was quicker and easier for me to recover from a mistake	1	2	3	4	5
11.	I was able to understand the concepts better with the microsoft office help function than I would have without them	1	2	3	4	5
12.	With the help of microsoft office help function I have managed to develop new abilities	1	2	3	4	5
13.	Microsoft office help function helped me to feel more confident about my computer skills	1	2	3	4	5
14.	I trusted the hint from microsoft office help function	1	2	3	4	5
Satisfaction levels about the microsoft office						
15.	The interactions with microsoft office were easy	1	2	3	4	5
16.	I was able to use microsoft office successfully	1	2	3	4	5
17.	I felt frustrated working with microsoft office	1	2	3	4	5
18.	I felt nervous when working with microsoft office	1	2	3	4	5
19.	I really had to concentrate to work with microsoft office	1	2	3	4	5
20.	It was exciting working with microsoft office	1	2	3	4	5
21.	Working with microsoft office made me change my attitude towards computers	1	2	3	4	5
22.	My experience with microsoft office encouraged me to learn about other computer programs	1	2	3	4	5
23.	Based on my experience with microsoft office, I can encourage my friends to learn about new concepts	1	2	3	4	5

24. What was the most difficult part when you worked with Microsoft Office? **(You may tick more than one option)**

Microsoft office environment was distracting
Microsoft office help function instructions were too difficult to follow
I understood very little from the microsoft office help function
Other, please specify:

---

25. What was the best part when you worked with Microsoft Office? **(You may tick more than one option)**

Easier to get information needed from microsoft office help function
Easy to understand and follow microsoft office help function instruction
Microsoft office help function instructions were straightforward
With Microsoft office help function it was easier to figure out how to perform a particular task
Other, please specify:

---

26. Did you enjoy being part of this study?

	YES	NO	

27. Provide reasons for your answer in question 26.

---

28. How much did you learn from the study overall?

A lot	sufficient	Average	Poor	Nothing
1	2	3	4	5

---

29. Any general comments or suggestions:

---

30. I would like to participate in a similar research project in future.

	YES	NO	

31. If you answered 'YES' in question 30, please provide your cell phone number:_____

Thank you very much for your input in this research.

Appendix B

Post-test Questionnaire (with agents)

The purpose of this questionnaire is to elicit your personal opinions of the agents you had worked with while carrying out different tasks for evaluation purposes. Please answer all the questions.

For question 1 to question 25 rate your opinion on a scale of 1-5, circling the appropriate number, where: 1 = "Strongly disagree", 2 = "Disagree", 3 = "Not sure", 4 = "Agree", 5 = "Strongly agree"

No.		SD	D	NS	A	SA
Learning contents provided by the agents						
1.	The agents had functions and capabilities I expected	1	2	3	4	5
2.	The agents used a language that was familiar to me	1	2	3	4	5
3.	The agents' hints provided all the necessary information	1	2	3	4	5
Effectiveness of the agents						
4.	The agents' hints helped me to quickly learn how to perform a particular task	1	2	3	4	5
5.	The agents' hints helped to recall the different steps involved for a particular task	1	2	3	4	5
6.	The agents' hints helped me to complete my tasks quicker	1	2	3	4	5
7.	The agents' hints were very practical	1	2	3	4	5
8.	The agents' hints assisted me to identify my mistakes when performing a task	1	2	3	4	5
9.	With the agents, it was quicker and easier for me to recover from a mistake	1	2	3	4	5
10.	I was able to understand the concepts better with the agents than I would have without them	1	2	3	4	5
11.	With the help of the agents I have managed to develop new abilities	1	2	3	4	5
Satisfaction levels about the agents						
12.	The interactions with the agents were easy	1	2	3	4	5
13.	I was able to use the agents successfully	1	2	3	4	5
14.	I trusted the advice from the agents	1	2	3	4	5
15.	I found the agents to be intelligent	1	2	3	4	5
16.	I found the agents to be friendly	1	2	3	4	5
17.	I felt frustrated working with the agents	1	2	3	4	5
18.	I felt nervous when working with the agents	1	2	3	4	5
19.	I really had to concentrate to work with the agents	1	2	3	4	5
20.	It was exciting working with the agents	1	2	3	4	5
21.	The agents' hints helped me to feel more confident about my computer skills	1	2	3	4	5
22.	Working with the agents made me change my attitude towards computers	1	2	3	4	5
23.	My experience with these agents encouraged me to find out more about them	1	2	3	4	5
24.	Based on my experience with the agents, I can encourage my friends to use them when learning about new concepts	1	2	3	4	5
25.	I would like to consider using agents when learning other concepts in real life	1	2	3	4	5

For question 26 to question 29 select whether the statement is true of false in terms of your preference

26.	I prefer a male agent to a female agent	True ☐	False ☐
27.	I prefer a cartoon agent to a realistic agent	True ☐	False ☐
28.	I prefer a dog agent to a human agent	True ☐	False ☐
29.	I prefer a text agent to a text and audio agent	True ☐	False ☐

For question 30 to question 32 select the one agent that you liked the most based on the criteria listed below

No.	*Liking levels of the agents' temperament*									
	Male text	Female text	Male Text and audio	Female Text and audio	Male cartoon dog	Female cartoon dog	Male cartoon human	Female cartoon human	Male realistic dog	Female realistic dog
30.	Appearance									
31.	Voice									
32.	Movement									

33. For each of these agents, indicate whether you enjoyed working with them or not. Please provide suggestions for improvement.

		Enjoyable	Frustrating	Suggestions
33.1	Male text agent			
33.2	Female text agent			
33.3	Male text and audio agent			
33.4	Female text and audio agent			
33.5	Male cartoon dog agent			
33.6	Female cartoon dog agent			
33.7	Male cartoon human agent			
33.8	Female cartoon human agent			
33.9	Male realistic dog agent			
33.10	Female realistic dog agent			

34. Select your first choice of agent in terms of your overall preference and indicate it with an 'X' (**select only one**)

34.1	Male text agent
34.2	Female text agent
34.3	Male text and audio agent
34.4	Female text and audio agent
34.5	Male cartoon dog agent
34.6	Female cartoon dog agent
34.7	Male cartoon human agent
34.8	Female cartoon human agent
34.9	Male realistic dog agent
34.10	Female realistic dog agent

35. Please give a brief reason for your 1st choice rating in question 34 (e.g. 1st choice agent was friendly, intelligent, attractive, etc.).

1st

_____

36. What was the most difficult part when you worked with agents? (**You may tick more than one option**)

_____

Agents were distracting
Agents were speaking too fast
I understood very little of what the agents said
Agents were saying the same things over and over again
Other, please specify:

_____

37. What was the best part when you worked with agents? (**You may tick more than one option**)

_____

Easier to get information needed
Easy to understand what the agents said
Agents' help and hints were straightforward
Easier to figure out how to perform a particular task
Other, please specify:

_____

38. Did you enjoy being part of this study?

	YES	NO	

39. Provide reasons for your answer in question 38.

_____

40. How much did you learn from the study overall?

A lot	sufficient	Average	Poor	Nothing
1	2	3	4	5

41. Any general comments or suggestions:

_____

_____

42. I would like to participate in a similar research project in future.

	YES	NO	

43. If you answered 'YES' in question 42, please provide your cell phone number: _____

Thank you very much for your input in this research.

References

1. Continental Corporation: ABET Against illiteracy in South Africa. http://www.contionline. com/generator/www/com/en/continental/csr/themes/society/education_science/abet_en.html
2. Lincicum, S.: Introduction to interface agents. http://www.ous.edu/onlinenw/2003/executive/ LincicumExecSumm.pdf
3. Potgieter, L.: Creating different virtual character representations (interface agents) in a simulated MS-WORD environment. BSc Honors Project. University of the Free State, South Africa (2010)
4. ISO 9241-11: Ergonomic requirements for office work with visual display terminals. Beuth, Berlin (1998)
5. Rudowsky, I.: Intelligent agents. In: Proceedings of the Americas Conference on Information Systems, NY (2004)
6. Giraffa, L.M.M., Viccari, R.M.: The use of agents techniques on intelligent tutoring systems. http://lsm.dei.uc.pt/ribie/docfiles/txt200342413856156.PDF (1998)
7. Bartneck, C., Croft, E., Kulic, D.: Measuring the anthropomorphism, animacy, likeability, perceived intelligence and perceived safety of robots. In: Proceedings of the Metrics for Human-Robot Interaction Workshop in affiliation. The 3rd ACM/IEEE International Conference on Human-Robot Interaction (HRI 2008), Technical Report 471, University of Hertfordshire, pp. 37–44. Amsterdam (2008)
8. Angeli, A.D.: Ethical implications of verbal disinhibition with conversational agents. PsychNology J **7**(1), 49–57 (2009)
9. Wonisch, D., Cooper, G.: Interface agents: preferred appearance characteristics based upon context. http://www.vhml.org/workshops/HF2002/papers/wonisch/wonisch.pdf (2002)
10. Chou, C.Y., Chan, T.W., Lin, C.J.: Redefining the learning companion: the past, present, and future of educational agents. http://chan.lst.ncu.edu.tw/publications/2003-Chou-rtl.pdf (2003)
11. Landowska, A.: The role and construction of educational agents in distance learning environments. In: Proceedings of the 1st International Conference on Information Technology. Gdansk 19–21 May 2008, pp. 321–324 (2008)
12. Technological Fluency Institute: Computer Literacy: What is computer Literacy & Why is it important. http://www.techfluency.org/computer-literacy.htm
13. Githens, R.P.: Older adults and e-learning. http://www.rodgithens.com/papers/older_ adults_elearning_2007.pdf
14. Nielsen, J.: Jakob Nielsen's Alertbox. Usability 101: Introduction to Usability. http:// www.useit.com/alertbox/20030825.html
15. Usability.gov: usability basics. http://www.usability.gov/basics/index.html
16. Peacock, M.: The what, why and how of usability testing. http://www.cmswire.com/cms/ web-engagement/the-what-why-and-how-of-usability-testing-007152.php
17. Adebesin, T.F., De Villiers, M.R., Semugabi, S.: Usability testing of e-learning: an approach incorporating co-discovery and think-aloud. http://researchspace.csir.co.za (2009)

18. Catrambone, R., Stasko, J., Xiao, J.: Anthropomorphic agents as user interface paradigm: experimental findings and a framework for research. In: Cognitive Science Society. The 24th Annual Conference of the Cognitive Science Society 2004, pp. 166–171 (2004)
19. Prendiger, H., Mori, J., Ishizuka, M.: Using human physiology to evaluate subtle expressivity of a virtual quizmaster in a mathematical game. Int. J. Hum. Comput. Stud. **62**(2), 231–245 (2004)
20. Bickmore, T., Cassel, J.: Social dialogue with embodied conversational agents. In: van Kuppevelt, J., Bernesen, N.O. (eds.) Advances in natural, multimodal dialogue systems. Kluwer, NY (2005)
21. Sahimi, S.M., Zain, F.M., Kamar, N.A.N., Samar, N., Rahman, Z.A., Majid, O., Atan, H., Fook, F.S., Luan, W.S.: The pedagogical agent in online learning: effects of the degree of realism on achievement in terms of gender. Contemp. Educ. Technol. **1**(2), 175–185 (2010)
22. Li, I., Forlizzi, J., Dey, A., Kiesler, S.: My agent as myself or another: effects on credibility and listening to advice. In: DPPI'07: Proceedings of the 2007 Conference on Designing Pleasurable Products and Interfaces, pp. 194–208. NY (2007)
23. Ratan, R., Bailenson, J.N.: Similarity and persuasion in immersive virtual reality. Panel presentation to the Communication and Technology Commission of ICA (2007)
24. White, M., Foster, M. E., Oberlander, J., Brown, A.: Using facial feedback to enhance turn-taking in a multimodal dialogue system. In: Proceedings of HCI International 2005 Thematic Session on Universal Access in Human–Computer Interaction (2005)
25. Kim, Y., Wei, Q.: The impact of learner attributes and learner choice in an agent-based environment. Comput. Educ. **56**(2011), 505–514 (2010)
26. Carmody, K., Berge, Z.: Pedagogical agents in online learning. http://ww.cogimedia.com/110revised.pdf (2008)
27. Steve in action. http://www.isi.edu/isd/VET/vet-body.html
28. Johnson, W.L.: Socially intelligent agent research at CARTE. http://www.aaai.org/Papers/Symposia/Fall/2000/FS-00-04/FS00-04-015.pdf (2000)
29. Sabot, A., Aini, Z.I., Lew, T.T.: Computer virus courseware using animated pedagogical agent. http://elib.unirazak.edu.my/staff-publications/iznora/computer%20virus.pdf (2005)
30. Bertrand, J., Babu, S.V., Polgreen, P., Segre, A.: Virtual agents based simulation for training healthcare workers in hand hygiene procedures. Lecture Notes in Computer Science, vol. 6356, pp. 125–131 (2010)
31. Foo, K.K.: Effects of pedagogical agents' instructional roles on learners with different cognitive styles in terms of achievement and motivation. Dissertation, University Sains Malaysia, Penang (2010)
32. Johnson, B., Christensen, L.: Educational research: quantitative, qualitative, and mixed approaches. SAGE, London (2012)
33. Department of Labour: Speech by Minister of labour, M. Mdladlana. http://www.info.gov.za/speeches/2004/04102015451003.htm
34. Investorwords. http://www.investorwords.com/3738/population.html
35. Kendra cherry: What is informed consent? http://psychology.about.com/od/iindex/g/def_informedcon.htm
36. Tullis, T., Albert, B.: Measuring the user experience: collecting, analyzing, and presenting usability metrics. Morgan Kaufmann, MA (2008)

Chapter 7
MASECO: A Multi-agent System for Evaluation and Classification of OERs and OCW Based on Quality Criteria

Gabriela Moise, Monica Vladoiu and Zoran Constantinescu

Abstract Finding effectively open educational resources and open courseware that are the most relevant and that have the best quality for a specific user's need, in a particular context, becomes more and more demanding. Hence, even though teachers and learners (enrolled students or self-learners as well) get to a greater extent support in finding the right educational resources, they still cannot rely on support for evaluating their quality and relevance, and, therefore, there is a stringent need for effective search and discovery tools that are able to locate high quality educational resources. We propose here a multi-agent system for evaluation and classification of open educational resources and open courseware (called MASECO) based on our socio-constructivist quality model. MASECO supports learners and instructors in their quest for the most appropriate educational resource that fulfills properly their educational needs in a given context. Faculty, educational institutions, developers, and quality assurance experts may also benefit from using it.

Keywords Evaluation and classification of open educational resources and open courseware · Multi-agent system for classification of open educational resources and open courseware · Socio-constructivist quality model for open educational resources and open courseware

G. Moise · M. Vladoiu (✉) · Z. Constantinescu
UPG University of Ploiesti, Ploiesti, Romania
e-mail: mvladoiu@upg-ploiesti.ro; monica@unde.ro; mvladoiu@yahoo.com

G. Moise
e-mail: gmoise@upg-ploiesti.ro

Z. Constantinescu
e-mail: zoran@upg-ploiesti.ro

M. Ivanović and L. C. Jain (eds.), *E-Learning Paradigms and Applications,*
Studies in Computational Intelligence 528, DOI: 10.1007/978-3-642-41965-2_7,
© Springer-Verlag Berlin Heidelberg 2014

7.1 Introduction

The Open Educational Resources (OERs) movement started in 2002 with the *Education Program of the Hewlett Foundation* introducing a key element into its strategic plan *Using Information Technology to Increase Access to High-Quality Educational Content*, which aimed at helping equalizing *the distribution of high quality knowledge and educational opportunities for individuals, faculty, and institutions* worldwide using the ICT support. The initial focus was twofold: funding *production* of exemplars of high-quality content and building community, collaboration, and a shared knowledge base about the *creation, dissemination, access, use and evaluation* of open educational resources [1].

The OER original model included, beside funding and promoting, living specifications of high-quality open content, establishing quality benchmarks for various forms of content, which have faded out in the more recent OER logic models. The desideratum of high-quality has been reached mainly by financing branded content from prestigious institutions. However, despite the strong arguments for this approach, it is crucial to *find additional mechanisms for vetting and enhancing educational objects in social settings, ways to close loops and converge to higher quality and more useful materials* [1]. While providing high-quality educational materials from top institutions will remain essential to the success of the OER Initiative (the spin-off open courseware movement rooted in the MIT OCW project is a prominent such successful example), the increasing role of the open repositories in this process is essential for creation of appropriate learning loops that continuously improve these materials through reflected use, re-use, re-mix etc. [1]. In this chapter, for the sake of easiness in phrasing, we use the acronym OCW for both OpenCourseWare and open courseware, where the former refers to projects based on the MIT OCW paradigm, while the latter regard any free offering of online courseware based on other paradigms.

One major challenge that the OER movement has had to face, due mainly to its significant success, has to do with the fast growing number of both open educational repositories and instructional resources available freely. Thus, finding effectively the resources that are the most relevant and that have the best quality becomes more and more demanding [2]. Hence, even though teachers and learners (enrolled students or self-learners as well) get more and more support in finding the right educational resources, they still cannot rely on support for evaluating their quality and relevance [3], and therefore there is a stringent need for effective search and discovery tools [2]. OPAL, one of the last Hewlett Foundation funded projects, has evaluated a wide range of international OER projects focusing on quality issues, and the conclusion was that *systematic quality assurance mechanisms for OER are lacking in higher education and adult education...* and that it is necessary to overcome the insecurity concerning how to validate the value of open educational resources and practices *when quality management approaches are largely absent for OERs* [4]. It is argued that OER providers themselves should be the first to ask for accreditation, certification, and quality assurance, so that their offerings comply with the standards in the field, and, therefore there will be more confidence in and acceptance of OERs [5].

The recent 2012 Paris OER Declaration, issued at the 2012 OER Congress in Paris, recommends that future support is needed for *facilitating finding, retrieving and sharing of OERs, for fostering awareness and use, for facilitating enabling environments for use of ICT, for reinforcing the development of strategies and policies, for promoting the understanding and use of open licensing frameworks, for supporting capacity building for the sustainable development of quality learning materials, for fostering strategic alliances, for encouraging the development and adaptation in a variety of languages and cultural contexts, for encouraging research, and for encouraging the open licensing of educational materials produced with public funds* [6].

In spite of the scale, pervasiveness, and influence of the growing movement of free sharing of educational resources and courseware on users around the world, there is yet no quality assessment framework that could provide support for (1) *learners* in their quest for finding the most appropriate educational resources for their educational needs in a given context, for (2) *instructors* who are interested in educational resources that support their teaching and learning activities, and provide for both achievement of learning goals, objectives, and outcomes, and for reflective learning, for (3) *faculty or institutions* that are or want to become involved in this movement, and they may be interested in the challenges and benefits of this process, for (4) *developers* who need guidelines for designing and building such educational resources, or for (5) *experts in quality assurance* of educational resources [7–9].

We propose here a multi-agent system for evaluation and classification of open courseware and open educational resources (called MASECO) based on our socio-constructivist quality model introduced in [7]. The main goal of MASECO is supporting OER/OCW users, being them learners, instructors, developers, evaluators, faculty, institutions, consortiums etc. to fulfill better their needs, and accomplish appropriately their educational aims, in any specific context. The criteria that constitute the backbone of the model have been grouped in four categories related with content, instructional design, technology, and courseware evaluation. Therefore, our work mainly supports two of the 2012 Paris OER Declaration recommendations that refer to support for both locating and retrieving OER that are relevant to specific needs (*facilitate finding, retrieving and sharing of OER*) and for promoting quality assurance of OERs (*sustainable development of quality learning materials*).

MASECO has three main components as follows: (1) an OER/OCW Management System, which is built on top of a database management system, and which manages both OERs and OCW (storing and updating information related to the OER and the OCW included in the system), (2) a *Classification Agent* that classify OERs and OCW using various classifiers, and (3) a *Communication Agent*, which manages the communication between agents and between the system and the environment.

Several use scenarios may take place according to the user's type. For example, for a regular user, the working scenario is as follows: the user interacts with the system through the Communication Agent, and she makes a request of the "best"

OER or OCW according to her needs, within a given context. The Communication Agent corroborates information from the OER/OCW Management System and sends it together with the user's needs to the OER/OCW Classification Agent, which returns a result to the Communication Agent. Finally, the Communication Agent transmits the system's answer back to the user. A developer may update and improve MASECO, so he interacts with it for updating both the OER/OCW database, and for initiation of training sessions for the neural network of the Classification Agent.

Our work is focused in the first place on helping learners, as they often do not have the background knowledge, information seeking or metacognitive skills necessary to evaluate effectively the educational value of digital resources, which is understandable taking into account that this capacity to evaluate and discriminate lays on the highest level of critical thinking skill in Bloom's taxonomy for thinking about educational goals [10, 11]. Nevertheless, all the other actors involved in educational processes mentioned before can benefit highly from using MASECO.

The structure of the chapter is as follows: the next section addresses issues of quality assurance for OERs and OCW as they are reflected in the literature, while the third one is concerned with the related work on concrete solutions for evaluation and classification of OER and OCW. Section 7.4 includes the research methodology. In Sect. 7.5, we give a detailed description of using MASECO for evaluation and classification of OERs and OCW with respect to the architecture, the conceptual models, the use scenarios, the used classifiers, the experimental results, and some discussion around the whole experience, the lessons learned, and the challenges that remain. The last section includes some conclusions, along with future work ideas.

7.2 Quality Assurance for OERs and OCW

Quality of open educational content is seen as a strong base for future sustainability of the OER/OCW movement, no matter what approach of ensuring quality is followed [1], given that movement's sustainability is essential to the successful, large scale OER/OCW acceptance, and embedding in education [12], being it at resource's production level or at resource's sharing level [13]. Quality Assurance (QA) is a frequently raised topic, as learning resources are expected to be trusted and authoritative [1, 3, 12–15]. Moreover, common lack of reviewing and quality assurance is seen as a critical issue that is holding back the increased uptake and usage of OCW and OERs [3, 12, 16]. Furthermore, ensuring quality of open content is seen as one of the major challenges to the growing OER/OCW movement, along with both lack of awareness regarding copyright issues and sustainability of the OER/OCW initiatives in the long run [3]. Other works see quality assurance as a key factor for sustainability, besides funding and support [12].

OER quality can be approached twofold: first, design processes that guarantee content quality and suitable formats, and secondly, development (or maintenance) processes to ensure currency [12]. Two quality assurance directions are followed:

the first one is concerned with discovering the way in which experts and users assess quality of digital resources in general, striving to identify both major factors influencing human judgments and lower level features of resources that people attend to when assessing quality. The second direction consists of applying machine learning algorithms when seeking solutions for this kind of problem. Of course, there are several issues to consider, such as including consistency (or not) of humans when making potentially subjective assessments and various features chosen to focus on when training machine learning algorithms. However, when asked to evaluate the quality of digital resources, people usually rely on a set of criteria (that often are implicit) to guide their reasoning [17]. In addition to sca-lability, another challenge of assessing educational resource quality is the matter of perspective: the quality is rather contextual than intrinsic to the resource. It depends on the configuration that corroborates the users' constituency, the edu-cational setting, and the intended educational purpose of the resource [17].

Moreover, traditional quality assurance mechanisms are not appropriate for assessing quality of OCW and OERs because of their high developing and changing rate. Easy to use, dynamic, and user-centered quality mechanisms are considered more suitable for ensuring quality of OCW and OERs through various community approaches: *peer reviews, user commenting, and rating* (not to exclude branding, of course). Self-sustainable, competent, and aware communities that work coordi-nately on resource creation and improvement, quality assurance, and experience sharing should be invigorated [3, 18, 19]. Open peer review according to a set of agreed criteria is seen as well suited for this task [20]. Other low-level elements that describe the use of a particular educational resource may be useful, such as the number of downloads, the argument for such an approach being that *quality is not an inherent part of a learning resource, but rather a contextual phenomenon*, and a learning resource maybe be or not useful in a specific learning situation, and therefore, the learner should be the judge of that [3, 5, 21]. It is argued that in spite of offering high quality materials and best pedagogical design and methods, OERs and OCW will prove their true value for learners only if they match the learners' own context, and therefore they are genuinely reusable (or at least fully adaptable) [22].

Over time, several solutions have been envisaged for coping with ensuring quality of OERs and OCW. For instance, some institution-based providers rely on the rep-utation (*brand*) of the institution to convince the learners that the offered open edu-cational materials are of high quality. Of course, most probably, the materials are subjected to some *internal quality assurance procedures* before being released as open content, but these are not open to the public so that they could be followed [3]. Likewise, a Global Index system has been proposed, which aimed at helping potential users to locate and access easily the needed courseware [16, 23]. It was supposed to be based on a *vetting mechanism* supported by a volunteer group acting as a de facto editorial board. More recent works sustain a similar idea of creation of *learning exchanges* that are focused directories linking to only high-quality repositories, and using only commonly established standards for classification and sharing [24].

Other approach suggests establishing formal co-operations between educational organizations that are involved in sharing and reusing of OERs from a common pool of content, tools and services, which it is thought as having a positive impact also on the quality leveraging of OERs, for example by being assessed critically by partner institutions, which may result in improved internal quality criteria and control [14, 15]. *Word-of-mouth* method is also seen as a viable quality management process [3].

Peer reviewing has been primarily used for quality assurance by some well-known open educational resources repositories, such as MERLOT (Multimedia Educational Resource for Learning and Online Teaching) or NLN Materials (National Learning Network Materials). All the MERLOT resources are assessed by discipline-specific editorial boards with regard to their quality of content, potential effectiveness as a teaching–learning tool, and their ease of use, while the NLN materials are peer reviewed by a range of colleges [4, 21, 25, 26]. In MERLOT, all the peer reviewers in a specific editorial board share and compare their evaluations to create test cases, which are used further on to develop evaluation guidelines that are to be applied to all the resources of each particular discipline [2].

However, the traditional peer reviewing process, which is focused on assessing the factual accuracy, intellectual content, and educational context of a resource, is not the most suitable for educational resources, as their educational quality is hard to evaluate outside the instructional context. Therefore, it is the educator's responsibility, when reusing a particular resource, to create suitable pedagogical scenarios within proper educational contexts [21]. MERLOT, for example, complements the formal peer review with recording user comments and ratings.

In [27] the authors analyze deeply the traditional peer review process, and after identifying the pressures this process has to face, and pointing out the fallacies of reviewing in the online world, they propose a general set of principles for understanding how peer review should be applied today to different kinds of content and in new platforms for managing quality. These principles consider not only the materials' content, but also their context of use, and while the focus here is on OERs, though they may be applied *across multiple levels of knowledge production, including scholarship and reference materials in addition to educational publishing* [27].

Context of use may indicate where a resource is being used by learners, such as in a classroom, in a laboratory, or as part of an encyclopedia, or it can describe the stage of a scientific work from the perspective of authors, such as draft, revised version, or updated version. It can also refer to contexts of reuse such as a translation, a derivative work for a different goal, or a constantly updated resource like a Wikipedia article. *The stress is on understanding the variety of contexts in which a resource exists and not only the end-user consumption of a resource* [27]. What resources are good in what contexts it is an outcome of the reviewing process.

The principles for review that authors propose are as follows: (1) *principle of maximum bootstrapping: designers of new systems should build on and adapt existing communities of expertise, existing norms for quality, and existing mechanisms of review;* (2) *principle of objectified evaluations: treat reviews as their own kind of object, disassociated from a single resource, specifying context of use, and potentially applicable to multiple versions;* (3) *principle of multiple magnifications: more reviews are better, more data about reviewers is better, because multiple, combined views on an object are now possible, with a corollary for the third principle: review is not blind, but pseudonymous and persistent.*

Another approach considers *a voluntary* (*or mix of voluntary and paid*) *wiki-like model, in which OERs are the object of micro-contributions from many. This approach raises complex issues of quality, but much work on collective "con-*verging to better" *is under way* [1]. Such an approach is taken in Connexions, where the traditional pre-publication review is replaced with a post-publication review based on a more open community of third party reviewers. Acknowledging that there are multiple perspectives on quality, Connexions permits third parties to use a mechanism, which provide different views onto collections [5, 17]. The mechanism that allows this process is based on the *lenses*. For example, a user, be it an individual, an institution or an organization may set up their own reviewing, then it is able to select the modules and collections that meet their quality standards, and, when the repository is accessed through that user's lens, only the materials they estimate as having the appropriate quality may be viewed. While Connexions users have access to all modules and courses in the Content Commons (whatever their stage of development and quality level), they also have the opportunity to preferentially locate and view modules and courses rated high quality by choosing from a range of different lenses provided by third parties, each lens having a different focus [1, 4, 28]. So, in this model, *pre-publication credentialed materials are not merely distributed through the network; post-publication materials are credentialed through use in the networks* [1].

Other approaches investigate the viability of generating dynamic use histories for educational resources, which will record each instance of using or reusing a particular resource. This will provide for searching the most used resources for a given topic [21], while other sees the quality assurance process as corroborating checking, peer reviewing, feedback, rating or voting or recommendation, and branding or provenance or reputation [29]. It remains to be seen which approach will gain acceptance within educational communities, but most probably that will combine some of the approaches presented briefly here and, in our opinion, it will have to do with user communities.

7.3 Related Work: QA and Classification of OER/OCW

After pursuing a very thorough search in various prestigious digital libraries and indexing databases, we have become aware that the related work is very scarce, with just a few works hardly similar with ours in some particular respects.

In [30] the author envisages various teaching and learning activities happening in a semantic web-based education environment, in which intelligent *pedagogical agents* provide the infrastructure needed for information and knowledge flows between user clients (authors, teachers, learners etc.) and educational servers. These agents are autonomous software entities that provide for human learning and cooperate with various actors involved in pedagogical processes and with each other, in the context of interactive learning environments. They assist searchers in locating, browsing, selecting, re-mixing, integrating, adapting, personalizing, re-using etc. educational materials located on different servers [30, 31].

Automatic identification of educational materials by classifying documents found on the web with respect to their educational value is explored in [32, 33]. The authors formulate the task as a text categorization problem, and prove that the generally accepted concept of a learning object's "educational value" can be reliably assigned through automatic classification by carrying out several experiments on a dataset of manually annotated documents, which show that the generally accepted notion of a learning object's "educational value" is a property that can be, in authors' view, reliably assigned through automatic classification. Furthermore, an examination of cross-topic and cross-domain portability illustrates that the *automatic classifier can be ported to other topics and domains, with minimal performance loss.* The authors have identified also several features of educational resources: the educational value, the relevance, the content categories (definition, example/use, questions and answers, illustration, other), and the resource type (class web page, encyclopedia, blog, mailing lists/forums, online book, presentation, publication, how-to article, reference manual, other). The expertise of the annotators is also retained as it is important when evaluating the educative value of a resource. The resources have been scored on a four point scale mapped to four labels: non-educational, marginally educational, educational, and strongly educational. Both papers present an experiment on a dataset of materials in Computer Science, while the second presents also another experiment on a dataset in Biology, to prove cross-domain portability.

Automatic classification of didactic functions of information objects, based on machine learning, aiming at increasing the re-use rate of digital learning resources, at various levels of granularity, is addressed in [34]. Each information object was manually labeled with its didactic function, according to Meder's didactic ontologies [35]. The function types have been hierarchically ordered on three levels as follows: the first one differentiates between receptive knowledge types and interactive assessments; the former is further divided into source, orientation (facts or overview), explanation (what-explanation or example), and action knowledge (checklist or principle), while the latter consists of either multiple choice tests or assignment tests. Nine features have been used to evaluate whether multimedia features can be used for classification of didactic functions. Four different classifiers were used and evaluated: a Bayes network classifier, a Support Vector Machine (SVM), a rule based learner and a decision tree learner. The classifiers worked on the three levels of details presented above, on a set of medical educational resources (training corpus of 166 information objects, further 207). Their results have been compared with the human evaluation (six evaluators). The main performance measure was classification accuracy, which has reached a level of over 70 % when hierarchical classification has been performed and a level of 85 %

when multi-label classification has been used. The authors have identified also two extra features that could improve these results: position of an information object within the learning object or course it belongs to (that may show author's intended learning strategy), and the style of speech (as it may vary and depends closely on particular didactic goals).

In [29] the authors consider four quality dimensions for OERs: content, pedagogical effectiveness, ease of use, and reusability, each one of them being detailed further on. The *content dimension* includes accuracy, currency, and relevance; the *pedagogical effectiveness* relies on learning objectives, prerequisites, learning design, learning styles, and assessment; the *ease of use* is related to clarity, visual attractiveness, engagement, clear navigation, and functionality, while *reusability* depends on format, localization, and metadata-based discoverability.

In [17] the authors worked on the idea that identifying concrete factors of quality for web-based educational resources, both used by experts in quality assessment and easily recognized by non-experts, can make manageable machine learning approaches to automatically determine quality characterization and educational value. The aim of their work was dual: empowering learners with tools for evaluation of quality of online educational resources and helping digital librarians to manage large educational collections. They were driven by the need to develop both methodologies able to identify dimensions of quality that are associated with specific educational goals and algorithms able to characterize resources with respect to these dimensions. Twelve dimensions of quality have been identified as the most important: good general set-up, appropriate pedagogical guidance, appropriate inclusion of graphics, readability of text, inclusion of hands-on activities, robust pedagogical support, age appropriateness, suitability of activities, connections to real-world applications, reflecting the source's authority, focus on key content, and access to relevant data. They constructed a training corpus of 1,000 digital resources annotated with these quality indicators, and trained machine learning models which were able to identify important indicators, with accuracies of over 80 %. The indicator extraction process has resulted in one numeric vector per educational resource in the digital library. The machine learning system (based on SVM) has analyzed the corresponding vectors for the training corpus, and it has learned a statistical model for some selected indicators. Further on, it has evaluated whether the quality indicators are present or not in a resource, based on applying those models to the vector corresponding to that particular resource.

In [11] the authors subscribe their work to the high goal of developing a computational model of quality that come close to human expert evaluations, based on machine learning and natural language processing, and, moreover, to provide automatic tools that implements that model. They started with performing an extensive literature review and meta-analysis, and, consequently identified 16 features of resources or metadata that could be useful for detecting quality variations across resources, lying in five categories: *provenance* (cognitive authority, site domain), *description* (element count, description length, metadata currency), *content* (resource currency, advertising, alignment to educational standards, word/link/image count on the first page of the resource, multimedia), *social authority* (Google's PageRank, annotations), and *availability* (cost, functionality). The experiment consisted in manual annotations for collections of DLESE digital library (600 resources), according to evaluators' *gestalt sense of quality* and personal preferences regarding quality. The collection rankings have been used as rankings for individual resources contained in the collections (to avoid the huge amount of effort necessary for manual annotations of each individual resource). Three classification categories have been used, namely A+, A, and A−, and each one was containing 200 resources. Within each classification category, the resources were further divided randomly into training and testing sets where 80 % of the resources were used to train the model and the remaining 20 % were used to test the accuracy of the trained model. The authors have computed metrics for evaluation of the quality indicators (only for the first page of each resource), and have experimented in a series of add-one-in analyses to see which indicators have positive, respective negative contributions to the classification. When all the indicators were used, their models were able to identify whether a resource was from a high (A+), medium (A), or low (A−) quality collection with 76.67 % accuracy, while when using only the quality indicators that positively contributed to the classification increased the models' accuracy to 81.67 %.

In [36] the authors propose a measure of relevance, which integrates a variety of existing quality indicators, and which can be automatically computed, building on work in [27, 37]. In [37], the author shows that the current systems for recommending educational materials lack a *weighting mechanism* that would allow assessment data from various sources to be considered. Consequently, he proposes an integrated quality indicator that combines explicit expert and user evaluations, anonymous evaluations and implicit indicators (such as favorites and retrievals). In [36] the authors reflect on Connexions, where each lens focuses the user's view on a subset of available modules and collections deemed high quality by the controlling authority, and propose combining lenses for filtering content. Therefore, they propose a relevance indicator that can be calculated automatically, as a sum of three weighted sums of quality indicators, which can be classified into three categories: *evaluative* that includes all explicit expert and user evaluations (overall rating, content quality, effectiveness, ease of use, comments), *empirical* that refers to information on materials usage, such as retrievals, the number of users who bookmark them, and so on (personal collections, exercises, used in classroom), and *characteristic* that refers to descriptive information about the materials, as obtained from their metadata (reusability). Thus, the explicit evaluations made by users or experts, the descriptive information obtained from metadata and the usage data are used in order to increase the reliability of recommendations by integrating various quality aspects.

In [38] the authors show that quality of learning objects may be improved by better educating their designers, by incorporative formative assessments and learning testing in design and development models, and by providing summative reviews that should be maintained as metadata, which users can use when searching, sorting, and selecting learning resources. They also point out the variety of settings in which OERs are produced and consumed, which results in needing more than one evaluation model. The authors present here also their instrument for reviewing learning objects (called LORI) that incorporates several aspects related to quality of such objects: content quality, learning goal alignment, feedback and adaptation, learners' motivation, presentation design, interaction usability, accessibility, reusability, and standards compliance. Furthermore, they use this instrument within a suite of tools for collaborative evaluation that small evaluation teams (including subject matter experts, learners, instructional designers) use to produce *an aggregated view of ratings and comments.* Adapted from LORI, in [39] seven rubrics are provided, five of them being adapted from LORI (*content quality, motivation, presentation design, usability, accessibility*), while the other two are new: educational value and overall rating. *Educational value* refers to its potential to provide learning on the addressed subject, to the accuracy, clarity, and unbiasedness of the information presentation, while the *overall evaluation* captures the perceived usefulness of resources in educational contexts.

An interesting approach is taken in [40] where the authors address very important issues when it comes to quality rating and recommendation of learning objects such as sharing evaluative data of learning objects across different repositories and combining various explicit and implicit measures of both quality and preference to make recommendations for appropriate learning objects that fulfill user's needs. In their endeavor, they had used Bayesian Belief Networks (BBNs), a powerful probabilistic knowledge representation and reasoning technique for partial beliefs under uncertainty. Moreover, BBN allow them to approach problems of insufficient and partial reviews in learning object repositories, as well as corroborating data from different quality evaluation instruments. They have been working with two learning object quality rating standards: MERLOT and LORI. First, they had produced a correlated structure between MERLOT Peer Review and LORI, and afterward they had constructed a BBN based on this structure, which has helped them to, for example, infer how a learning object would be rated on MERLOT's ease-of-use item, given actual ratings on LORI's interaction-usability and accessibility items. They present real-world BBNs that have been constructed to *probabilistically model relationships among different roles of reviewers (both expert and anonymous), among various explicit and implicit ratings, and among items of different evaluation measurements,* along with the results of a qualitative study and of simulated test cases. Their BBNs are able to derive *the implications of observed events, the rated attributes, by propagating revised probabilities throughout the network, when each attribute's value is updated.* Based on their experience they conclude that *the BBN model makes quantitatively reliable inferences about different dimensions of learning object quality* and that *the availability and accuracy of quality ratings can be largely improved in a learning object repository.*

Finally, Achieve in collaboration with leaders from the OER community have developed a rubric, aiming at helping various actors involved in education to determine the degree of alignment of OERs to the Common Core State Standards and to determine quality aspects of OERs [41]. Recently, Achieve has teamed up with OER Commons to develop an online evaluation tool based on that rubric, and currently, OER Commons hosts both the tool and its resulting assessment data [42]. Each resource of OER Commons may be evaluated, the resulting information is stored in a pool of metadata, and it may be shared through the Learning Registry with other interested repositories [43]. The Achieve rubrics includes the following components: degree of alignment to standards, quality of explanation of the subject matter, utility of materials designed to support teaching, quality of assessment, quality of technological interactivity, quality of instructional and practice exercises, opportunities for deeper learning, and assurance of accessibility.

7.4 The Research Methodology

During the initial stage of this work we have been searching for OCW and/or OERs that cover the necessary content for an introductory course on databases. We had performed several thorough searches in various repositories, such as MIT OpenCourseWare, OCW Consortium, Saylor Foundation, University of Washington Computer Science and Engineering courses, Coursera, OER Commons, Webcast. Berkeley, Connexions, Universia OCW, ParisTech, Open.Michigan, University of California, Irvine, University of Southern Queensland, Utah State University, Intute, Textbook search [44–57], and much more others. We have been using either the repository's specific search capabilities, or "classic" Google searches. Furthermore, we have exploited Google's custom OER/OCW search and particular OCW search engines alike [58, 59]. Our first goal has been the identification of as many possible candidates for our further research on quality assessment. The wanted candidates have consisted of "full" online open courseware and/or educational resources that provided support for a course on database fundamentals (being it OpenCourseWare or any other kind of complete courseware—even as a proper mix of OERs—available freely online).

Despite our best efforts we have ended up with just eight viable candidates for our further work, this being due to a variety of reasons, for instance some open courseware was available only in some foreign languages we could not understand, or others consisted only in video recordings of actual teaching of the course content in the classroom. The finalists are eight open courseware that offer educational materials on databases [60–67], provided by various open courseware repositories that comply with different open courseware paradigms, as follows (each one of them has been assigned an acronym, for easier further presentation and discussions):

- the MIT OpenCourseWare on *Database Systems*—1-MIT-OCWDB;
- the Saylor Foundation's *Introduction to Modern Database Systems* open courseware – 2-Saylor-DB;
- the Stanford's Professor Jennifer Widom *Introduction to Databases* open courseware—3-St-WidDB;

- the Introduction to Database Systems courseware provided by Nguyen Kim Anh in Connexions—4-Cnx-NKA;
- the King Fahd University's KFUPM OpenCourseWare on *Database Systems*— 5-KF-DBSs;
- the University of Washington's Introduction to Data Management open courseware—6-UW-DMg[344];
- the Universidad Carlos III de Madrid's *Database Fundamentals* (*Fundamentos de las bases de datos*) OpenCourseWare—7-UC3M;
- the Universidad Politecnica de Madrid's Database Administration (*Administracion de bases de datos*) OpenCourseWare—8-UPM-BD.

To score the candidates we have used our own rubric based on our quality assurance criteria for open courseware and open educational resources that builds up on our previous work in [68, 69], which it was introduced in [7], put to work in [8, 9], and refined further for this work. These criteria correspond to the quality characteristics of *quality in use, internal and external product quality* according to ISO/IEC 25000 SQuaRE standard, and they cover the next user needs: effectiveness, efficiency, satisfaction, reliability, security, context coverage, learnability, and accessibility. These quality criteria may be used for quality assessment of either small learning units or an entire courseware. They have been grouped in four categories related with *content, instructional design, technology* and *courseware evaluation*, which will be briefly explained further on.

Content related: This category includes criteria that reflect whether the resource provides the online learners with multiple ways of both engaging with their learning experiences and achieving of the content's mastery. First criterion refers to the easiness of using the resource, reflected by *readability* and *uniformity of language, terminology, and notations*. Another useful element is the *availability of the course syllabus*, so that users become aware since the very beginning of the content scope and sequence. The *comprehensiveness of the lecture notes*, i.e. whether the course content and assignments demonstrate sufficient wideness, deepness and rigor to reach the standards being addressed, is also to be retained in our quality model. *Modularity of the course content* is also important, as modular course components are units of content that may be distributed and accessed independently, giving each user both the *possibility to select the most suitable learning unit* at a particular time and the *opportunity to choose the most appropriate learning path* that matches the user's needs and abilities. The course materials may be approached easily *top-down, bottom-up*, or in a *combined* way. *Availability of assignments* (with or without solutions), being them exercises, projects, and activities, is important as well, as they are content items that enhance the primary content presentation. When looking at a particular learning resource, other than an entire courseware, which can be a small learning unit, a course module, a lesson etc., users are particularly interested in various characteristics of the resource: *accuracy, reasonableness, self-containedness, context, relevance, availability of multimedia inserts, correlation of the resource with the course in its entirety, links to related readings*, and *links to other resources* (*audio, video* etc.).

Instructional design related: criteria address the instructional design and other pedagogical aspects of teaching and learning for that resource. They include the educational resource's *goal and learning objectives*, which are expected to be clearly stated and measurable, as the learner's level of knowledge mastery and practical abilities is to be measured against both the main goal and each and every learning objective. The educational materials are ought to provide for multiple opportunities for learners to be actively engaged in the learning process, having meaningful and authentic learning experiences during undertaking various *appropriate instructional activities*: problem- or project-based learning, e-simulations, learning games, webcasts, scavenger hunts, guided analysis, guided research, discovery learning, collaborative learning groups, case studies etc. *Learning outcomes* state the learner's achievements after performing a learning activity, i.e. what learners will know and/or will be able to do as a result of such an activity, in terms of knowledge, skills, and attitudes. The *availability of the evaluation and auto-evaluation means* (with or without solutions) is also important from a pedagogical point of view. The teacher users may be also interested in the *learning theory* (behaviorist, cognitivist, constructivist, humanist and motivational etc.) and in the *instructional design model* (ADDIE, ARCS, ASSURE etc.) that have been used to develop that particular educational resource. Moreover, learning experiences that provide for *reflective learning* will always add to the overall quality of educational resources. Under the reflection perspective, the desired outcome of education becomes the construction of coherent functional knowledge structures adaptable to further lifelong learning. Reflection has a dual sense here: one would be the process by which an experience, in the form of thought, feeling or action is brought into consideration (while is happening or subsequently), and the other refers to the creation of meaning and conceptualization from experience and to the potentiality to look at things from another perspective (critical reflection) [70–73].

Technology related: Both open educational resources and open courseware are expected to benefit fully from ICT technologies, to have user-friendly interfaces, to comply with standards for *interoperability*, and to provide for appropriate access for learners with special needs (*accessibility*). *Extensibility* of each educational resource, aiming at expanding learning opportunities, from a technological point of view, refers to easiness of adding content, activities, and assessments both for developers and learners. A high quality *user interface* is based on technical aspects related to the capabilities of the supporting hardware, software and networking. A clear specification of the technology *requirements* at user's end (both hardware and software), along with the *prerequisite skills* to use that technology are useful to help users understand how the resource should be used to benefit fully from its content. A high quality open educational resource is expected to work smoothly on a variety of platforms in use around the world (*multi-platform*). Having a true engaged learning relies on learner's opportunity to interact with the content and with other learners, which is not possible without a suite of rich *supporting tools*.

Courseware evaluation: Despite the initial claim of just offering high quality educational materials to learners worldwide, with no other intention to support learners during their learning journeys, all major open courseware initiatives have started to be more involved with their learners. Hence, regular assessment of effectiveness of open courseware becomes essential, along with using the results for further improvements. Each prospective user would most probably first be interested in the *courseware overview*, which includes information about the *content scope and sequence, the intended audience, the grade level, the periodicity of updating the content, the author's credentials and the source credibility, its availability in multiple-languages, instructor facilitation or some kind of semi-automated support, suitableness for self-study and/or classroom-based study and/ or peer collaborative study, the time requirements, the grading policy,* along with *instructions about using* that courseware and its components, in order to establish the most suitable learning paths, the reliability, and the availability of links to other educational resources (readings, OCW, OER etc.). *Prerequisite knowledge* and *required competencies* are also useful for to be known by users at the beginning of a learning process. *Matching the course schedule*, if any, with learner's own pace, is also desirable. Another useful criterion regards the *terms of use (service)*, i.e. availability of repository or institutional policies with respect to copyright and licensing issues, security for primary, secondary and indirect users, anonymity, updating and deleting personally identifiable information, age restrictions, neti-quette, etc. OERs and OCW that are *free of bias and advertising* are also desirable. Suitable design and presentation of educational content is also considered, along with *user interface richness (style)* as it is defined by its navigational consistency, friendliness, multimedia inserts, interactivity, adaptability (both to user's needs and context) etc. Another quality criterion is concerned with the option to provide, or aiming to provide, a formal degree or a certificate of completion (*degree or certificate*). *Participatory culture and Web* 2.0 facets are also important being them related to contribution to the content, collection of users' feedback, collab-oration with fellow teachers/learners/developers and so on, or to sharing the development or using experience.

To sum up, we have evaluated each resource's quality using a number of 69 criteria that are presented briefly in Table 7.1 (our rubric). The fulfillment of each criterion has been assessed on a scale between 0 and 5, where the scoring meaning has been as follows: 0 = absence, 1 = poor, 2 = satisfactory, 3 = good, 4 = very good and 5 = excellent. The assessment has been performed indepen-dently by three evaluators having between 10 and 20 years of experience with teaching both fundamentals of and advanced databases for undergraduate and graduate students. For the next step of the process, which will be presented in the following section, we have used a "negotiated value" around the arithmetic mean of the scores, which has resulted from a panel reviewing process that involved the three reviewers.

Table 7.1 Criteria for quality assurance of OCW and OER

Content related	*To what degree an OER/OCW allows learners to have **engaging learning experiences** that provide for **mastery of the content***	
	• CR1: readability	0–5
	• CR2: uniformity of language, terminology, and notations	0–5
	• CR3: availability of the course syllabus	0–5
	• CR4: comprehensiveness of the lecture notes	0–5
	• CR5: modularity of the course content	0–5
	• CR6: possibility to select the most suitable learning unit	0–5
	• CR7: opportunity to choose the most appropriate learning path	0–5
	• CR8: top–down, bottom–up or combined approach	0–5
	• CR9: availability of assignments (with or without solutions)	0–5
	• CR10: *resource related*: accuracy[1], reasonableness[2], self-containedness[3], context[4], relevance[5], multimedia inserts[6], interactive elements[7], correlation with the entire course[8], links to related readings[9], links to other resources (audio, video etc.)[10]	$0\text{–}5 \times 10$
Instructional design	*Criteria that address the **instructional design**, and other **pedagogical aspects** of Teaching and Learning (T&L) for that resource*	
	• ID1: goal and learning objectives (**outline** the material)	0–5 (1 global + 4 per unit)
	• ID2: learning outcomes (students will know/be able to do—**skills, abilities, attitudes**)	0–5 (1 global + 4 per unit)
	• ID3: appropriate instructional activities	0–5
	• ID4: availability of the evaluation and auto-evaluation means (with solutions)	0–5 (ex./ others$(1 + 1.5) \times 2$)
	• ID5: learning theory	0–5
	• ID6: instructional design model	0–5
	• ID7: *reflective learning opportunities* in which the desired outcome of education becomes the construction of coherent functional knowledge structures adaptable to further lifelong learning	0–5
Technology related	*Both OERs and O CW are expected to **benefit fully from ICT technologies**, and to comply with various standards*	
	• TR1: conformity with standards for interoperability	0–5
	• TR2: compliance with standards for accessibility	0–5
	• TR3: *extensibility*: easiness of adding content, activities and assessments, from a technological point of view (both developers and learners)	0–5 (2.5 + 2.5)
	• TR4: user interface's basic technological aspects (hw-device, sw., networking)	0–5
	• TR5: supporting technology requirements at user's end	0–5
	• TR6: prerequisite skills to use the supporting technology	0–5
	• TR7: multi-platform capability	0–5
	• TR8: supporting tools	0–5

(continued)

Table 7.1 (continued)

Courseware evaluation	Despite of the original claim of just offering high quality educational materials, all major open courseware initiatives have recently become more **involved with their learners**. Hence, regular assessment of **effectiveness of open courseware** becomes essential, along with using the results for further improvements	
	• CW1: *courseware overview*: content scope[1] and sequence[2], intended audience[3], grade level[4], periodicity[5] of content updating, author's credentials[6], source credibility[7], multiple-languages[8], instructor facilitation[9] or semi-automated support[10], suitableness for self-study[11], classroom-based[12] study, and/or peer collaborative[13] study, time requirements[14], grading policy[15], instructions on using[16] the courseware, reliability[17], links to other[18] educational resources (readings, OCW, OERs etc.)	0–5 × 18
	• CW2: availability of prerequisite knowledge	0–5
	• CW3: availability of required competencies	0–5
	• CW4: matching the course schedule with learner's own pace	0–5
	• CW5: *terms of use (service)*: availability of repository or institutional policies wrt copyright and licensing issues, security for primary, secondary and indirect users, anonymity, updating and deleting personally identifiable information, age restrictions, netiquette, etc.	0–5
	• CW6: freeness of bias and advertising	0–5
	• CW7: suitable design and presentation of educational content	0–5
	• CW8: *user interface richness (style)*: navigational consistency[1], friendliness[2], multimedia[3], interactivity[4], adaptability[5] (both to user's needs and context) etc.	0–5 × 5
	• CW9: providing a formal degree or a certificate of completion	0–5
	• CW10: *participatory culture and Web 2.0 facets*: contribution to the content[1], collection of users' feedback[2], collaboration with fellows[3], sharing the development[4]/using[5] experience	0–5 × 5

7.5 Using MASECO for QA and Classification of OERs and OCW

In this section we will present briefly our approach of evaluating and classifying open courseware (that can be used also for open educational resources) with help from a multiagent system. We started with the working definition from [74] that states that *an agent is a computational mechanism that exhibits a high degree of autonomy, performing actions in its environment based on information (sensors, feedback) received from the environment*. A multi-agent environment includes more agents, which interact with one another, and further, they have to work under constraints of the environment not knowing constantly and continuously every-thing about the world that other agents know. These constraints are essential to the definition of a real multi-agent system [74].

MASECO is a multi-agent system whose main goal is to register and classify OERs and OCW, based on a quality model. The architecture of the system contains three main components: two intelligent agents (the Communication Agent and the Classification Agent), and the OER/OCW management system. The classification process aims to assign objects to predefined categories. Most automatic classifi-cation endeavors are grounded in the area of machine learning, which describe algorithms that learn behavior (e.g. how to classify an object) based on training information [34, 75, 76]. Typical methods are Support Vector Machines (SVM), decision tree learners, Bayes classifiers, and artificial neural networks. Such algorithms work as follows: they start with a training corpus of objects, for which the category is known. After a training phase new objects can be classified as well. Of course, classifiers do not take complete objects as input, and rely on mapping each object to a set of features. In our case, this set consists of the 69 scores obtained for the quality criteria presented in the previous section, and our Clas-sification Agent uses artificial neural networks to perform this task.

The two agents have been built using the BDI (Beliefs, Desires, Intentions) approach [77, 78]: the informational, motivational, and deliberative states of an agent are described by means of beliefs, goals, plans, and intentions. Each of the agents is based on the INTERRAP architecture [79], which defines an agent as having three layers: a behavior-based layer, a local planning layer, and a cooperative planning layer, which allow the agent to combine reactive and deliberative reasoning, and to interact with other agents or with the environment. The INTERRAP architecture was also used in the iLearning system [80], and the promising results obtained there have determined us to further use the same architecture in MASECO.

There are two general usages of the term "agent" [81]: weak and strong. The weak notion denotes a hardware or software system having the following prop-erties: autonomy, social ability, reactivity, and pro-activeness. The stronger notion of an agent is used by the AI researchers to describe systems that exhibit, in addition to the above mentioned properties, concepts that are mostly applied to humans, such as knowledge, belief, intention, and obligation.

Generally, a multi-agent system is considered to be a network of multiple intelligent agents, which are interacting with each other, with and within an environment, in order to solve problems otherwise difficult (or even impossible) to be solved only by one agent. Together they can combine different intelligent techniques to attain superior performance, either from a computational point of view, or with respect to the complexity of the interaction between them. Multi-agent systems can be considered as a distributed artificial intelligence, empha-sizing the joint cooperation of agents with their own behavior and autonomy.

The main characteristics of a multi-agent system are as follows [82]:

- each agent has an incomplete view, with incomplete information and limited capabilities for solving the main problem;
- there is no global control of the system;
- the data is decentralized;
- the computation is asynchronous.

There are many learning methods that can be used in multi-agent systems. The choice of the learning method depends on the given problem, and it can be sometimes a very difficult task. Standard supervised, unsupervised, and reinforcement learning techniques can be used as starting points. *Supervised learning* requires the existence of an expert to provide a set of training examples (training data set). Each example is a *pair* consisting of an input object and a desired output value (target). By analyzing the training data set, the algorithm produces either a classifier (discrete output) or a regression (continuous output). *Unsupervised learning* does not require the existence of an expert. It tries to find the hidden structure in unlabeled data, so that there is no error when evaluating a potential solution. Many methods used for unsupervised learning are from data mining. *Reinforcement learning* is a method where the system learns from the continuous interaction with the environment. The goal of the agent is to collect as much reward as possible, so it can use its past experience, and it can choose any necessary action. MASECO uses two learning techniques: supervised learning and reinforcement learning. Supervised learning is used for the training of the Classification Agent, while reinforcement learning is used by both the Communication Agent and the Classification Agent.

The agents of MASECO have the following properties:

- *Reactivity*—the agents maintain a permanent connection to the environment and they adapt to its changes, like for example, the appearance of a new OER or OCW;
- *Interactivity*—the agents collaborate in order to reach the system's objective;
- *Autonomy*—the agents know when and how to initiate the required actions;
- *Proactivity*—the agents have explicit goals and objectives;
- *Instruction*—the agents use automated learning techniques.

7.5.1 The INTERRAP Architecture

The INTERRAP architecture was proposed by Jörg Müller [79]. The model is a layered, hybrid BDI model, with three layers that describe an agent:

- a *behavior based layer* incorporating reactivity and procedural knowledge for routine tasks;

- a *local planning layer* that provides facilities for means-ends reasoning for achievement of local tasks and for producing goal-directed behavior;
- a *cooperative planning layer* that enables agents to reason about other agents and that supports coordinated action with other agents.

Beliefs are split into *a world model*—containing object-level beliefs about the environment, *a mental model*—holding meta-level beliefs about the environment, and a *social model*—holding meta-level beliefs about other agents [78, 79]. Specific situations, namely relevant subsets of the agent's beliefs, trigger the initiation of actions. Situations are abstract representations of classes of world states, which are of interest for an agent. The three classes of beliefs correspond with three classes of situations as follows: *bevioral situations* that are a subset of the world model, *local planning situations*, which description is based on both world and mental models, and *cooperative planning situations* that contain in addition parts of the social model. Accordingly, the agent's goals can be reaction goals, local goals, and cooperative goals. *The operational primitives enable the agent to do means-end reasoning about how to achieve certain goals. They include patterns of behaviors and joint plans* [79].

7.5.2 The Architecture of MASECO

MASECO is a multi-agent system, which includes both intelligent agents and an OER/OCW Management System (built on top of a database management system). For the time being, there are two types of agents: a Communication Agent and Classification Agent, which, despite having different goals, they collaborate with each other. The Communication Agent, besides its role as a communication facilitator internally between the components of the system, and externally between the system and the environment, acts as a supervisor and coordinates the whole working scenario. The database management system contains all the information regarding any known OER/OCW to the system. The Classification Agent knows how to classify any of the OER/OCW, and it collaborates with the Communication Agent in order to obtain all the necessary information it needs to perform this task. This agent is intelligent, reactive, and task oriented. The conceptual model of MASECO is presented in Fig. 7.1. The Communication Agent can communicate with the environment, with real users, or with other existing Multi-Agent Systems (MAS).

MASECO use scenarios are presented in Figs. 7.2 (bird eye's view) and 7.3 (more detailed).

MASECO interacts with the environment using the Communication Agent. Through the agent's sensors, the system receives data, requests, and commands, and it sends them to the agent's control unit. The control unit of the Communication Agent decides the use type (1, 2 or 3). In the first case, a simple querying of the OER/OCW management system is performed, and the classification of the

Fig. 7.1 MASECO—the conceptual model

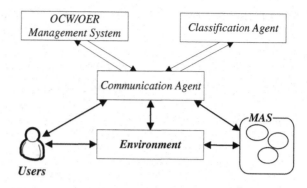

Fig. 7.2 MASECO—use scenarios (bird eye's view)

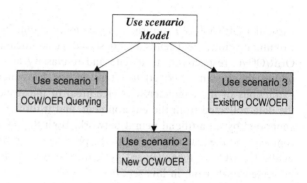

Fig. 7.3 MASECO—use scenarios, in detail

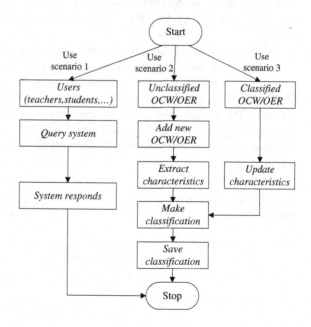

Fig. 7.4 The OER/OCW
data model

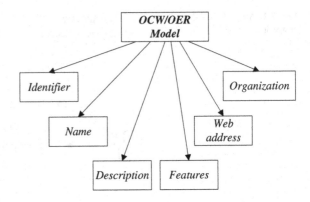

particular OER/OCW is returned to the environment. In the second two cases, the Communication Agent sends a request to the Classification Agent to classify a new OER/OCW, respectively to update and re-classify an OER/OCW already stored within the system. The control unit of the Classification Agent initiates the extractions of the characteristics (features) of the respective resource, processes the data collected from the environment, and it further on classifies the resource, supported by an artificial neural network. Both the data extracted from the environment and the data resulted from the processing are stored, respectively updated in the OER/OCW management system. The models for the two agents will be presented further on, in this section.

The ontology of MASECO is defined mainly by the OER/OCW model (Fig. 7.4), the use model (Fig. 7.2), and the classification instrument model (Fig. 7.5).

As we said previously here, the Communication Agent has the role of supervisor of MASECO. Its architecture is based on the agent model from the INTERRAP system, and it is shown in Fig. 7.6. The *world model* includes knowledge about OCW and OERs, and the agent updates information about OCW and OERs in the database. The *mental model* includes knowledge about itself, and its capacity to solve certain tasks. The *social model* includes knowledge about the Classification Agent, its capacities, action times, classification algorithms, etc. The situations that

Fig. 7.5 The classification
methods' model

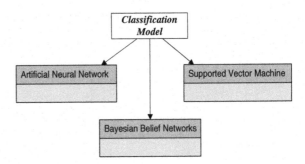

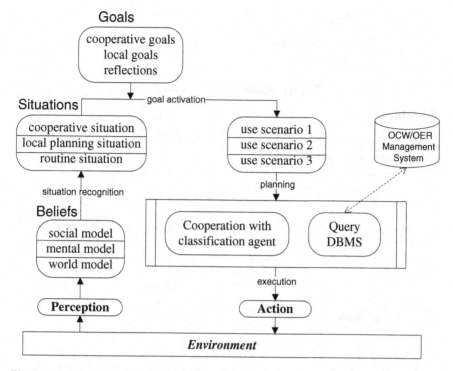

Fig. 7.6 The Communication Agent model

are recognized by the Communication Agent are as follows: *routine situations* (for instance to respond to a human user at a OER/OCW classification request and to justify that classification) that have a reflection goal, *local planning situations*, namely procedures for feature extraction (even by interacting with a human user) that have a local goal, and cooperative situations that include the cooperation with the Classification Agent for performing a classification, which have a cooperative goal.

The Classification Agent is simpler, and it is described in Fig. 7.7. It contains only two layers: a local planning layer and a cooperative layer. The local planning layer contains a classification plan, based on a classification algorithm. The cooperation layer is necessary to obtain the required information about the OER/OCW resource for the classification step. The Classification Agent has the ability to learn how to classify an open educational resource or open courseware. At the time being, MASECO uses for classification an artificial neural network. We have also tried other classifier and those results will be presented in a further subsection.

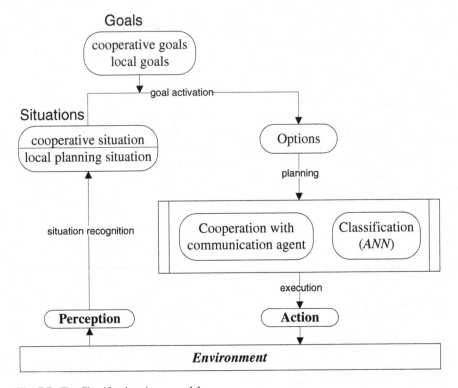

Fig. 7.7 The Classification Agent model

7.5.3 How MASECO Classifies OERs/OCW Using Artificial Neural Networks

Multi-class pattern classification refers to the problem of finding a correspondence between a set of inputs (that represents some characteristics) and a set of outputs (that represents two or more pattern classes). The classification relies on a variety of classifiers: feed-forward artificial neural networks, supported vector machine, decision trees, Bayesian belief networks, rule-based etc. A classification system usually has two components: a feature extractor and a class selector [83]. The architecture of MASECO's OER/OCW classification sub-system, which is a part of the Classification Agent, it is shown in Fig. 7.8.

Any OER/OCW that is to be classified will be pre-processed and its feature set will be extracted and stored in the corresponding feature vector, which will be the input for the classifier. The label of the obtained class is randomly checked and the results are interpreted by the learning strategies of the Classification Agent. The procedures of this agent decide on the classifier, which is, in fact, the kernel of any classification system. This may be modified by the actors of the Classification Agent, resulting in the selection of another classifier, or in a change of its structure.

Fig. 7.8 The Classification
Sub-system, a part of the
Classification Agent

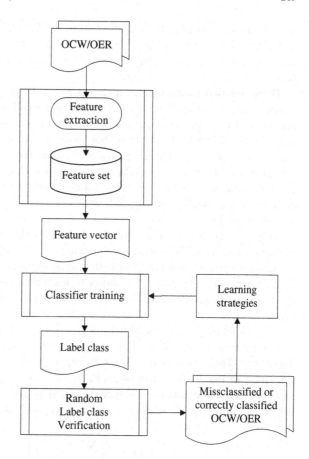

For example, the need for a new class of OER/OCW may determine the change of
the neural network structure, e.g. the increasing of the neuron number on the
output layer (as this number equals the number of classes).

The two processes, feature extraction and class selector can be formalized as
follows: [83, 84]: feature extraction is defined as a transformation $X = \Phi(P)$,
where $P = (p_1, p_2, \ldots, p_m)$ represents the pattern vector that describes an object, and
$X = (x_1, x_2, \ldots, x_d)$ is the feature vector (m is the number of object characteristics
and d is the dimension of the feature space. Using the vector X, the class selector
chooses a class $c_i \in C$, where $C = \{c_1, c_2, \ldots, c_k\}$ is a set of classes (k is the number
of classes). In our case, the process of feature extraction has been done using human
experts, while the class selector uses artificial neural networks. Considering that our
goal has been to classify OER/OCW resources in more than two classes, we had used
a multi-class neural network.

A multi-class neural network classification problem can be formalized as follows [85]. Having a d-dimensional feature space Ω with all the vector elements Y,
and a set of training data $\Omega_{training} \subset \Omega$, for each element from the $\Omega_{training}$ set we

consider as associated a class label cl from the *Class_labels* $= \{cl^1, cl^2, \ldots, cl^k\}$ set, where $cl^i \neq cl^j$, for all $i \neq j$ and $k > 2$. A classification system (F) based on artificial neural networks can be trained on $\Omega_{training}$ such that for each feature vector $Y \in \Omega$, $F(Y) \in$ *Class_labels*.

There are two major system architectures, a single artificial neural network system with m outputs that are determined by the codification scheme for pattern classes, and a system consisting of m artificial neural networks (binary artificial neural networks with a single output node or artificial neural networks with multiple output nodes). Three types of approaches for modeling pattern classes are available: one-against-one (OAO), one-against-all (OAA) and P-against-Q (PAQ). The experimental results show that an architecture with only one neural network performs well when the training data set is not too large and the pattern classes are not too many [85].

Artificial neural networks were introduced in 1943 by Warren McCulloch and Walter Pitts in [86]. These structures inspired from biology, from the neural circuit of nervous systems, are composed of interconnected computing units. In 1958, Frank Rosenblatt introduced the ability to learn and, consequently, he developed the perceptron model [87]. The model of an artificial neuron is presented in Fig. 7.9.

Each input of each artificial neuron has associated a synaptic weight w. This weight determines the effect of the corresponding input to the activation level of the neuron. The weighted sum of all the inputs $\sum w_j i_j$, with $j = 1 \ldots d$, defines the activation of the artificial neuron and it is called *net input*. The f function represents the activation function or specific function, and θ represents the threshold value. The output o of the neuron is computed using the following formula:

$$o = f\left(\sum_{j=1}^{d} w_j i_j - \theta\right) \tag{7.1}$$

The activation function f can have any of the possible forms presented in Table 7.2. In its simplest form, this is a binary function: either the neuron is firing or not, and it is described mathematically by the step function. In this case a large number of neurons must be used in computation beyond linear separation of categories. Other, more complex functions are also possible for the activation

Fig. 7.9 Model of an artificial neuron

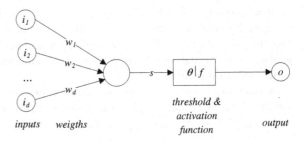

Table 7.2 Activation functions for artificial neural networks

Step function	$f(s) = \begin{cases} 0, s \leq 0 \\ 1, s > 0 \end{cases}$
Signum function	$f(s) = \begin{cases} -1, s \leq 0 \\ 1, s > 0 \end{cases}$
Linear function	$f(s) = s$
Sigmoid function	$f(s) = \frac{1}{1+e^{-ks}}, k > 0$
Generalized sigmoid function	$f(s) = \frac{1}{1+a*e^{-bs}}, b > 0$
Other interval-linear functions	$f(s) = \begin{cases} -1, & s \leq -1 \\ s, & -1 < s < 1 \\ 1, & s \geq 1 \end{cases}$
	$f(s) = \begin{cases} 0 & s \leq 0 \\ s, & 0 < s < 1 \\ 1, & s \geq 1 \end{cases}$

function. The *nonlinearity* of the activation function allows networks of neurons to compute nontrivial problems using a smaller number of nodes (Table 7.2).

More details about artificial neural networks and their applications can be found in [88]. We have also experimented with them before to adapt the teaching and learning process to the needs of learners within e-learning systems [89].

One of the first artificial neural networks was the feedforward neural network. In this case, the units (neurons) are connected in such a way not to form a directed cycle and the information flows from the input to the output, in only one direction, through any (if existing) hidden nodes. The simplest form is the single-layer perceptron network, which consists of only one layer, the output nodes. The inputs are fed directly to the output layer. A *perceptron* often refers to networks consisting of just one neuron. However, a single layer network is quite limited in its computational power. Multiple layers of such computational units (neurons), interconnected in a feedforward way, form the Multi-Layer Perceptron (MLP).

In MASECO, for the k-class classification of OERs and OCW, we have chosen a feedforward multi-layer artificial neural network (MLP), which is shown in Fig. 7.10.

The multi-layer perceptron has the following structure:

- *One input layer*—with d units of input that transmit the signal they receive;
- *One output layer*—with k units of output, with one neuron for each classification category;
- *One hidden layer*—with h units of neurons, which receive the information from the input layer and process it;
- *Connections*—each input neuron is connected to all the neurons from the hidden layer, so that we have complete connectivity; in a similar way, all the neurons from the hidden layer are forward connected to all the neurons from the output layer; each connection has an associated weight factor, w_{ij}, with $i = 1...d$, $j = 1...h$ for the input-hidden layers connectivity, and v_{ij}, $i = 1...h$,

Fig. 7.10 MASECO—
k-class classifier (MLP)

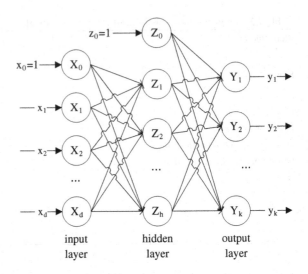

$$z_0 = 1 \longrightarrow Z_0$$

input layer   hidden layer   output layer

$j = 1\ldots k$ for the hidden-output layers connectivity; these weights can be easily represented as two matrices: $W = (w_{ij})$, $i = 1\ldots d$, $j = 1\ldots h$, and $V = (v_{ij})$, $i = 1\ldots h$, $j = 1\ldots k$, respectively;

- *Bias*—X_0 and Z_0 are used to define a threshold for the activation of the neurons.

In MASECO OERs and OCW can be classified in three categories: Satisfactory, Good, and Very Good. To define the pattern recognition problem, we had to arrange the characteristics of each resource as a column of dimension 69 (one value for each feature) in a matrix p, and similarly the target vectors in a matrix t with columns of dimension 3, representing the classification categories as shown below:

- (1, 0, 0)—satisfactory
- (0, 1, 0)—good
- (0, 0,1)—very good.

The tuple (p, t) defines a pattern.

The data set contains 140 input vectors of 69 elements each, one element for each characteristic and 140 target vectors of 3 elements each. It contains information extracted from the analysis of eight courseware and trivial data, with very low or very high characteristics. The numerical data is presented as such in Annex 1.

To perform our experiments we had used the Neural Network Pattern Recognition Tool of MATLAB. The generated network has been a feedforward network with one input layer, one hidden layer, and one output layer. On the input layer there are 69 neurons, while on the output layer only 3 neurons are present, which correspond to the three classes. The transfer functions for each layer are sigmoid. The number of neurons on the hidden layer is 10, which provides a neural network with good performances. The performance of the neural network might be improved by increasing the number of neurons in the hidden layer. The proposed network is shown in Fig. 7.11.

Multi-layer networks use different learning techniques. In our case, we have been using back-propagation. The output values are compared with the correct answer by computing a predefined error-function. In our case, the Classification Agent learns based on (p, t) tuples, namely pairs of OER/OCW and the corresponding class. Using some techniques, the error is fed back through the network and the algorithm adjusts the weights of each connection with the goal of minimizing the error function. After a large number of training cycles, the system will converge to a small error of the calculated function, so that we can consider that the network has "learned" the target function. Our network has been trained using scaled conjugate gradient backpropagation (*trainscg*).

We have divided the data set into three categories: training, validation, and testing. This has been done randomly. From the available 140 data samples available, we had used 70 % for training, 15 % for validation, and 15 % for testing (see Fig. 7.12). The results of the training are shown in Fig. 7.13.

The confusion matrix shows the percentages of correct and incorrect classifications. In general, a confusion matrix is a symmetrical array of the number of classified data compared to the actual data (the truth). The diagonal values represent the percentage of correctly classified data in each class. Correct classifications are the green squares (light gray in black and white) on the matrices diagonal, while

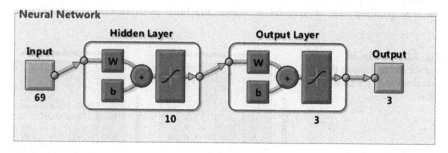

Fig. 7.11 MASECO—the neural network for OER/OCW classification

Fig. 7.12 Data set selection for training, validation, and testing

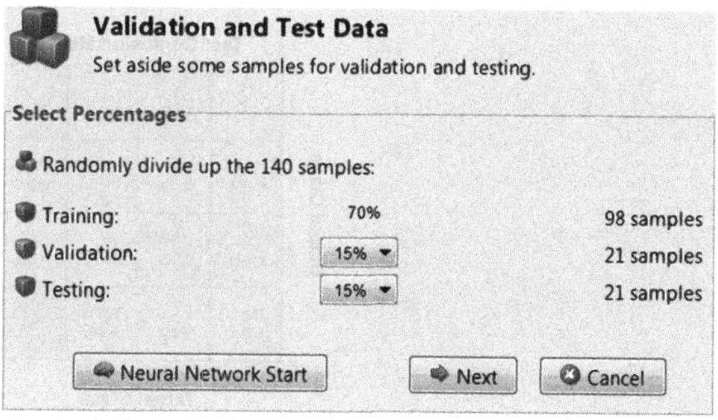

Fig. 7.13 Network training
net_OCW

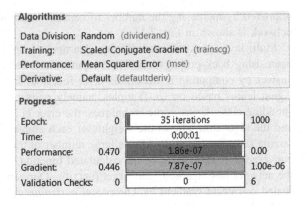

incorrect classifications form the red squares (medium gray in black and white). The blue squares represent the overall accuracies (the bottom right corner in all four matrices). If the network has learned to classify properly, the percentages for the incorrect classifications should be very small, indicating few misclassifications. The confusion matrices for training, testing, and validation data of our experiments are presented in Fig. 7.14, where we can see that the overall accuracies of All Confusion Matrix is high (99.3 %).

The Receiver Operating Characteristic (ROC), or ROC curve, is a graphical plot illustrating the performance of a binary classifier system. It is created by plotting the true positive rate or sensitivity versus the false positive rate (1-specificity),

Fig. 7.14 Confusion
matrices from training the
neural network for OER/
OCW

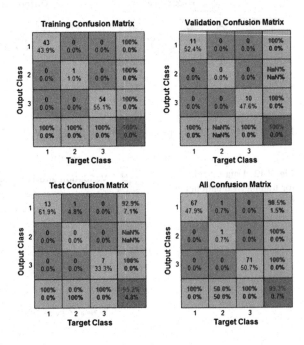

at various threshold levels. The ROC curve shows the functionality of the network. The best possible prediction method would show points in the upper-left corner, with 100 % sensitivity and 100 % specificity, i.e. the point (0, 1). This point is also called a *perfect classification*. A completely random guess would give a point along a diagonal line. In our case, the classification, as shown by the ROC curve, is very good as closest to the top border and to the left one the curve is, the classification is better (Fig. 7.15).

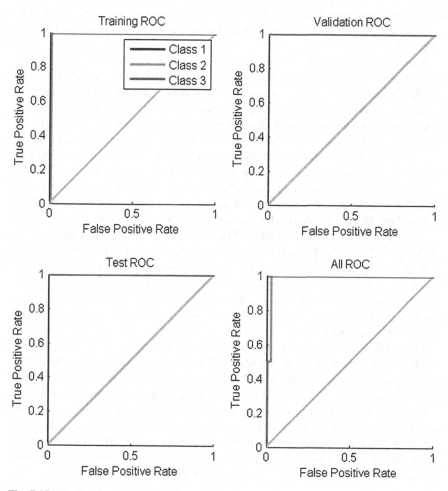

Fig. 7.15 The Receiver Operating Characteristic (*ROC*) curve

7.5.4 Classification of OERs and OCW Using Bayesian Belief Networks

In this sub-section we will overview briefly our experience about trying to classify OERs and OCW using Bayesian Belief Networks (BBNs), which are a very potent model for probabilistic knowledge representation and reasoning for partial beliefs under uncertainty [40, 90]. Uncertainty may refer to, for example, insufficient knowledge. In our case, this can refer to a partial quality evaluation that does not include all the 69 criteria or to the situation in which there are not enough assessments for some resources. BBNs combine two powerful theories that concern graphs and probabilities and provide for representing and updating beliefs (probabilities) about events of interest, as the quality score of an OER or OCW in this case.

Moreover, they allow performing probabilistic inference, for example to infer a quality score. From a mathematical point of view BBNs are directed acyclic graphs *in which the nodes represent propositional variables of interest (for example a feature of an objector the occurrence of an event), and the links represent informational or causal dependencies among the variables, which are quantified by conditional probabilities for each node given its parents in the network* [90]. Therefore, the probability of any subset of variables may be calculated given evidence about any other subset, and the reasoning process can work by propagating information in any direction. However, Bayesian networks are direct representations of the world and not of the reasoning process [90]. BBNs rely on both Bayes' Theorem (that has been introduced by Thomas Bayes, and it has been further explained by Richard Price in the sense that he has expressed the philosophical basis of Bayesian statistics [91]) and Bayesian probability theory with its core propagation mechanism. *The real power comes when we apply the above theorem to propagate consistently the impact of evidence on the probabilities of uncertain outcomes in a BBN, which will derive all the implications of the beliefs that are input to it. They are usually the facts that can be checked against observations* [90].

Our purpose here has been, once again, to predict accurately the class of each OER or OCW based on its quality scores. During our classification we have been using the following probability model for the classifier based on Bayes's Theorem [92, 93] (see Eq. 7.2). This classifier learns the class-condition probabilities $P(F_i = f_i | C = c_l)$ of each variable F_i in the data set (the quality scores), $i = 1\ldots69$, given the class label c_l. A new test case ($F_1 = f_1, \ldots, F_{69} = f_{69}$) is next classified based on Bayes's Theorem to compute the posterior probability of each class c_l given the vector of observed (evaluated) variable values (where C is the class variable and f_i refers to each possible value of F_i).

$$P(C = c_l | F_1 = f_1, \ldots, F_{69} = f_{69}) = \frac{P(C = c_l)P(F_1 = f_1, \ldots, F_{69} = f_{69} | C = c_l)}{P(F_1 = f_1, \ldots, F_{69} = f_{69})}$$

$$(7.2)$$

Fig. 7.16 The naïve BBN
used for classification

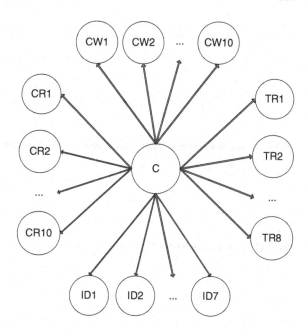

In our first attempt, we had used naïve Bayesian networks that characterize the situation in which the features that determine the membership to a class are independent, each attribute having a unique parent [94] (see Fig. 7.16).

Our first attempt to classify with the above naïve BBN has failed as it can be seen in the next screenshot (Fig. 7.17). This is due to the fact that naïve BBN treats by default all features as being a part of a normal distribution, and, therefore, it cannot work with a column that has zero variance for all the features related to a single class, which is, in fact, the case for some of the criteria and classes in our test. The reason is that there is no way for a naïve BBN to find the parameters of the probability distribution by fitting a normal distribution to the features of that specific class. To "force" a classification, we have altered insignificantly the scores that had this problem and, after training (Fig. 7.18), we had obtained successful classifications for several tests (one of them is shown in Fig. 7.19—a classification of a resource having very high scores as very good, i.e. in the class 3).

7.5.5 Discussion

We have experienced here with our multi-agent system the evaluation and classification of OERs and OCW based on a quality model with 69 quality criteria. To see whether such educational resources can be classified automatically has been our main goal. From our experience we have learned that this can be done provided that the right classifiers are used. Our first attempt that had used artificial neural network has been successful for our data set, while the second one that had

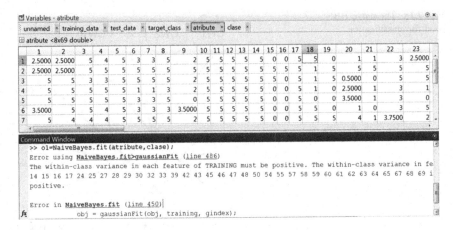

Fig. 7.17 Failed naïve BBN-based classification

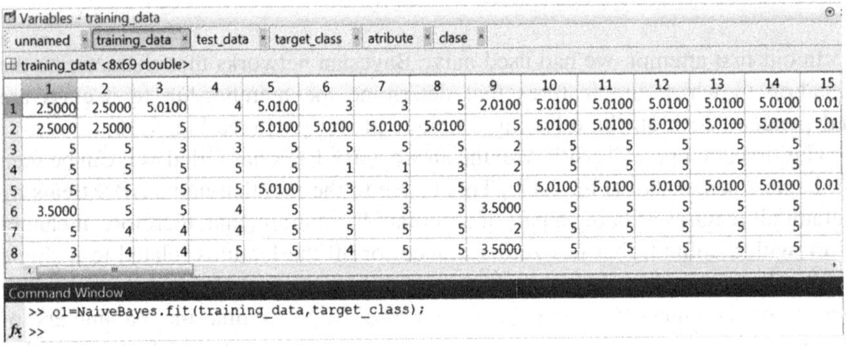

Fig. 7.18 The training of our naïve BBN

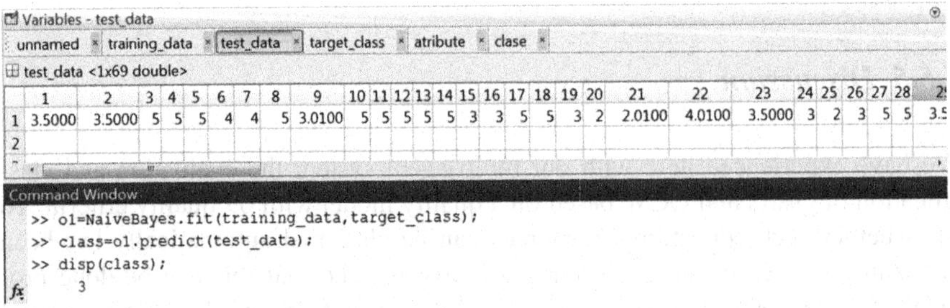

Fig. 7.19 Successful naïve BBN-based classification

used naïve Bayes networks has had some problems as we show in the previous sub-section. Of course, we consider as future work taking into account the intrinsic dependences that exist between some of the quality scores (features). For instance, the quality model scores the availability of assignments (with or without solutions) as a content-related criterion and the availability of evaluation and auto-evaluation means (with solutions) as an instructional-design related one. Of course, the evaluation in the two cases is capturing different aspects: in the first case, it refers to what the resource has to offer in terms of providing for engaging learning experiences that contribute to the mastery of the content (and having assignments with solutions may be very helpful in that respect), whilst in the second case, it reveals the importance of having your level of knowledge tested and self-tested. Another example refers to the aspects related to the user interface. First, we analyze it from a technological point of view with respect to the hardware, software and networking capabilities, and, secondly, we evaluate its richness (style), but once again, one cannot have a rich interface based on poor technological means. So the next step in this direction would be to determine the actual dependencies that exist between the quality criteria and, consequently, between the obtained scores. Based on those dependencies, in our future trials of classification of OERs and OCW using Bayesian Networks, we intend to use *Tree Augmented Naive Bayes* (*TAN*), which outperforms the naive Bayes classification, yet at the same time preserves the computational simplicity and robustness that characterize the naive Bayes method [92]. Of course, other classifiers are also envisaged: completely unrestricted Bayes classifiers, decision trees, rule-based, SVM etc., along with comparisons and cross-validation of the results. Furthermore, our current neural network is static, and we need to experiment whether a dynamic one that changes its structure to include new characteristics, new categories etc. would serve better our purposes.

Moreover, we are aware that we have to extend significantly both the set of evaluated resources and the pool of reviewers. Currently, we are in course of gathering together various OCW and OERs (around 10 resources per subject) that are necessary to graduate majoring in Computer Science, in a common repository of resources. Further on, we intend to have all the collected resources evaluated against the quality model by as many reviewers as possible so that we obtain a significantly larger amount of data to work on, using various classifiers. Some of these operations will be performed automatically by MASECO's agents.

Other issue to be considered concern the "contributing problem", i.e. how to convince as many reviewers as possible to perform quality assessment using our quality model. Quality reviews are not common because evaluating the quality of educational resources takes time, effort, and expertise [40]. We plan to develop a rubric-applying tool that facilitates human assessment so that the evaluators become keen to perform it. We also consider evaluating automatically some of the criteria, which can be learned by parsing intelligently each resource website. Human evaluators may keep these automatic results or they may change them to reflect their viewpoint. This could help also with incomplete evaluations that have scores only for some of the quality criteria. To obtain assessments from learners' point of view we think to involve our undergraduate and graduate students in making evaluation for their semester projects.

A weighting mechanism between the assessments of various users could be also useful to favor, for instance, a subject-matter expert's or a instructional designer's evaluation when compared with one of an anonymous on-line user. False positive (unfair) evaluations should be banned somehow. Especially when the number of quality assessments for a particular resource is low, the danger of altering the real quality resulted from evaluation is high in case of unfair assessment. BBNs seem to be helpful again in these situations as they are able to reduce the negative impact of distorted rating to a minimum degree [40].

Another direction to work on is concerned with objective measurements that could be included in the quality model: number of accesses, time spent with a resource, number of bookmarks, number of times a bookmark is followed, number of citations etc. Nevertheless, the semantics of such information has to be modeled properly within the quality model, if ought to complement seamlessly the explicit quality ratings.

7.6 Conclusions and Future Work

The open education movement has the potential to change the education world to a status quo of increasing richness and diversity, where educational resources, teaching and learning styles, and the huge variety of educational content can be tailored to more specific user needs and contexts. The ability to approach and solve quality assurance issues is a key aspect of this movement. The capacity to maintain the openness of the growing number of open education projects worldwide, and to further innovate on ways to guide the improvement of quality of open educational resources and open courseware through cooperation and collaboration, may open up new ways for education that are able to *match the complexity of the contemporary world and the many challenges it faces* [17].

Computational models of quality and automated approaches for computing the quality of digital educational resources will most probably be a part of the next generation of cognitive tools aimed at supporting users in making quality decisions. Therefore, ascertaining useful quality indicators and developing algorithms for automatically computing quality metrics and classifying resources based on these indicators are important steps towards reaching this goal. Concerns about the quality of educational resources found in digital libraries and repositories often revolve around issues of accuracy of content, suitability to the intended audience, appropriate design and information presentation, and completeness of associated descriptions (metadata and others) [11]. Having modeled and developed the proper suite of tools for assessing quality, we may imagine future educational infrastructures that support various users of educational resources in several quality evaluation processes. The cost of developing new educational resources may be also reduced by providing reliable quality assurance mechanisms that can support users in finding, using, and reusing high-quality open courseware and open educational resources [14]. The OER/OCW movement has also benefits from an individual point of view as well, as open sharing is claimed to increase publicity, reputation, and altruism of sharing with peers.

While the traditional view of quality assurance of educational content is seen as the responsibility of subject and instructional experts, in the context of OCW, OERs, and Web 2.0, guaranteeing quality seems more and more a community endeavor based on the collaboration between experts in education, subject scholars, students, teachers, developers etc. both during and after the teaching and learning process through study groups and practice communities around the world [95]. The emergent competition among OER/OCW initiatives calls for establishing of strong brands, of vivant user communities, and of improved quality of both resources and infrastructures [3].

Open sharing of OERs and OCW provides for broader and faster dissemination of knowledge, and thus ever more people are involved in problem solving, which in turn leads to rapid quality improvement and faster decentralized technical and scientific development. Therefore, *free sharing of software, scientific results and educational resources reinforces societal development and diminishes social inequality* [3]. This way, the OER Initiative's initial goal of *building a community so that the emerging OER movement will create incentives for a diverse set of institutional stakeholders to enlarge and sustain this new culture of contribution* may be reached [1].

We introduced here a multi-agent system (called MASECO) for evaluation and classification of open courseware and open educational resources, which is based on our socio-constructivist quality model, and which aims to support OER/OCW users to fulfill better their needs, and to accomplish appropriately their educational aims, in any given context. Our future work will research various issues related to quality evaluation of open educational resources and open courseware, both automatic and within communities of users. Some of these ideas have already been mentioned in the Discussion sub-section. Moreover, to disseminate this work further, one of the first things to do consists of creating a project wiki, as a starting point to build a community of users that could help with evaluating the materials, aiming at developing an educational repository that includes links to the most valuable educational resources for specific teaching and learning needs in various contexts. In this perspective, we consider the distributed management of information among the agents, as opposed to the centralized approach taken currently. We think also about proving out our approach cross-discipline and cross-domain with help from a case-based recommender system. Another idea we would like to pursue refers to refining our quality model towards a hierarchical approach, aiming at categorizing open educational resources for specific contextual needs (for example, most suitable for classroom study or for self-study).

Acknowledgments The authors are very grateful to both the editors and the anonymous reviewers for their valuable comments and ideas to improve this chapter.

Appendix: The Quality Scores Obtained by the Eight Open Courseware on Databases

	1 MIT OCWDB	2 Saylor DB	3 St WidDB	4 Cnx NKA	5 KF DBSs	6 UW DMg344	7 UC3M DADB	8 UPM BD
CR1	2.5	2.5	5	5	5	3.5	5	3
CR2	2.5	2.5	5	5	5	5	4	4
CR3	5	5	3	5	5	5	4	4
CR4	4	5	3	5	5	4	4	5
CR5	5	5	5	5	5	5	5	5
CR6	3	5	5	1	3	3	5	4
CR7	3	5	5	1	3	3	5	5
CR8	5	5	5	3	5	3	5	5
CR9	2	5	2	2	0	3.5	2	3.5
CR10.1	5	5	5	5	5	5	5	5
CR10.2	5	5	5	5	5	5	5	5
CR10.3	5	5	5	5	5	5	5	5
CR10.4	5	5	5	5	5	5	5	5
CR10.5	5	5	5	5	5	5	5	5
CR10.6	0	5	5	0	0	0	0	0
CR10.7	0	5	0	0	0	0	0	0
CR10.8	5	5	5	5	5	5	5	5
CR10.9	5	1	1	1	0	5	5	1
CR10.10	0	5	5	0	0	0	5	0
ID1	1	5	0.5	2.5	3.5	1	4	4
ID2	1	5	0	1	1	0	1	1
ID3	3	5	5	3	3	3	3.75	3
ID4	2.5	5	5	1	0	5	2	2.5
ID5	0	0	0	0	0	0	0	0
ID6	0	0	0	0	0	0	0	0
ID7	0	5	5	0	0	1	0	0
TR1	5	5	5	5	5	5	5	5
TR2	5	5	5	5	5	5	5	5
TR3	2.5	2.5	2.5	4	2.5	2.5	2.5	2.5
TR4	2	3	2	5	2	2	2	2
TR5	5	5	5	5	5	0	0	0
TR6	5	5	5	5	5	5	5	5
TR7	5	5	5	5	5	5	5	5
TR8	5	5	5	5	0	5	2	0
CW1.1	4	5	4	2	5	4	5	4
CW1.2	4	5	4	0	5	4	5	4
CW1.3	5	4	0	0	0	0	1	3
CW1.4	5	4	0	0	5	0	0	3
CW1.5	0	0	0	0	0	0	0	0
CW1.6	5	2.5	5	2.5	4.75	4.75	2.5	2.5

(continued)

(continued)

	1 MIT OCWDB	2 Saylor DB	3 St WidDB	4 Cnx NKA	5 KF DBSs	6 UW DMg344	7 UC3M DADB	8 UPM BD
CW1.7	5	5	5	3	3	5	5	5
CW1.8	0	0	0	0	0	0	0	0
CW1.9	0	0	2	0	0	0	0	0
CW1.10	0	2	2	0	0	2	0	0
CW1.11	5	5	5	5	5	5	5	5
CW1.12	5	5	5	5	5	5	5	5
CW1.13	0	5	5	0	0	0	0	0
CW1.14	0	5	0	0	0	0	0	0
CW1.15	2	5	5	2	2	2	0	2
CW1.16	4	5	5	0	0	0	0	0
CW1.17	5	5	5	4	3	5	5	5
CW1.18	1	5	1	1	0	5	2	2
CW2	5	5	5	0	0	0	5	0
CW3	5	5	5	0	0	0	0	0
CW4	5	5	5	5	5	5	5	5
CW5	5	5	5	5	5	5	5	0
CW6	5	5	5	5	5	5	5	5
CW7	2	5	5	2	2	2	2	2
CW8.1	5	5	5	5	5	5	5	5
CW8.2	2	5	4	3.75	2	2	2	2
CW8.3	0	5	3	0	0	0	0	0
CW8.4	0	5	1	0	0	0	0	0
CW8.5	2	5	2	5	2	2	2	2
CW9	0	5	5	0	0	0	0	0
CW10.1	0	2	0	3	0	0	0	0
CW10.2	2	5	5	5	2	0	0	0
CW10.3	0	3	3	0	0	0	0	0
CW10.4	0	3	5	4	0	0	0	0
CW10.5	0	3	5	4	0	0	0	0

References

1. Atkins D., Seely Brown J., Hammonds A.: A review of the open educational resources (OER) movement: achievements, challenges, and new opportunities. www.hewlett.org/uploads/files/Hewlett_OER_report.pdf (2007)
2. OECD: Giving knowledge for free—the emergence of open educational resources (2007)
3. Hylén, J.: Open educational resources: opportunities and challenges, pp. 49–63, Utah State University, Logan, UT. www.oecd.org/edu/ceri/37351085.pdf (2006)
4. Kernohan, D., Thomas A: Open educational resources—a historical perspective. http://repository.jisc.ac.uk/4915/
5. Albright, P.: Discussion highlights, pp. 61-83. In: D'Antoni, S. (ed.) OERs—conversations in cyberspace, UNESCO, Paris. www.col.org/SiteCollectionDocuments/country…/OER_Full_Book.pdf (2009)

6. UNESCO: Paris OER declaration. http://www.unesco.org/pv_obj_cache/pv_obj_id_EEF3C7E6694B8B91C31F5EA3340D484EF03A0100/filename/Paris%20OER%20Declaration_01.pdf (2012)

7. Vlădoiu, M.: Quality criteria for open courseware and open educational resources. In: 11th International Conference on Web based Learning 2012 (ICWL 2012), Workshops—2nd International Symposium on Knowledge Management and E-Learning (KMEL 2012), LNCS 7697, Sinaia, Romania (2012)

8. Vladoiu, M., Constantinescu, Z.: Evaluation and comparison of three open courseware based on quality criteria. In: Grossniklaus, M., Wimmer, M. (eds.) 12th International Conference on Web Engineering 2012 (ICWE 2012) Workshops—3rd Workshop on Quality in Web Engineering 2012 (QWE 2012), LNCS vol. 7703, pp. 204–215. Springer, Heidelberg (2012)

9. Vlădoiu, M.: Towards assessing quality of open courseware. In: 11th International Conference on Web Based Learning 2012 (ICWL 2012) Workshops—2nd International Workshop on Creative Collaboration through Supportive Technologies in Education (CCSTED 2012), LNCS 7697, Sinaia, Romania (2012)

10. Bloom, B.S., Englehart, M.D., Furst, E.J., Hill, W.H., Krathwohl, D.R.: Taxonomy of Educational Objectives: The Classification of Educational Goals. David McKay, New York (1956)

11. Custard, M., Sumner, T.: Using machine learning to support quality judgments. D-Lib Mag. 11(10), 1082. http://www.dlib.org/dlib/october05/custard/10custard.html

12. Nikoi, S., Armellini, A.: The OER mix in higher education: purpose, process, product, and policy. Distance Educ. 33(2), 165–184 (2012)

13. Wiley, D.: On the sustainability of open educational resource initiatives in higher education, OECD-CERI. www.oecd.org/edu/ceri/38645447.pdf (2007)

14. Geser, G.: Open educational practices and resources. OLCOS Roadmap 2012 Recommendations, Salzburg Research, EduMedia Group. Salzburg. http://www.olcos.org/cms/upload/docs/olcos_roadmap_recommendations.pdf (2007)

15. Geser, G.: Open educational practices and resources. OLCOS Roadmap 2012, Revista de Universidad y Sociedad del Conocimiento 4(1), 4–13 (2007)

16. Downes, S.: Models for sustainable open educational resources. Interdisc. J. Knowl. Learn. Objects 3, 29–44 (www.oecd.org/edu/ceri/36781698.pdf) (2007)

17. Bethard, S., Wetzer, P., Butcher, K., Martin, J. H., Sumner, T.: Automatically characterizing resource quality for educational digital libraries. In: 9th ACM/IEEE-CS Joint Conference on Digital Libraries, Austin, TX, USA (2009)

18. Pawlowski, J.M., Hoel, T.: Towards a global policy for open educational resources: the Paris OER Declaration and its implications, White Paper, Version 0.2, Jyväskylä, Finland (2012)

19. Larsen, K., Vincent-Lancrin, S.: The impact of the ICT on tertiary education: advances and promises. In: Kahin, B., Foray, D. (eds.) Advancing Knowledge and the Knowledge Economy. MIT Press, Cambridge (2006)

20. Taylor, P.: Quality and web-based learning objects: towards a more constructive dialogue, in Quality Conversations. In: 25th HERDSA Annual Conference, Perth, Western Australia, pp. 655–662 (2002)

21. Littlejohn, A.: Reusing Online Resources: A Sustainable Approach to E-Learning. Routledge, London (2003)

22. Richter, T., McPherson, M.: Open educational resources: education for the world. Distance Educ. 33(2), 201–219 (2012)

23. UNESCO: Forum on the impact of open courseware for higher education in developing countries (final report), Paris. http://www.wcet.info/resources/publications/unescofinalreport.pdf (2002)

24. Camilleri, A.F., Ehlers, U.D., Conole, G.: Mainstreaming OEP—recommendations for policy, OPAL Consortium. http://www.oer-quality.org/publications/project-deliverables/ (2011)

25. MERLOT: http://www.merlot.org

26. NLN Materials: www.nln.ac.uk

27. Kelty, C.M., Burrus, C.S., Baraniuk, R.G.: Peer review anew: three principles and a case study in postpublication quality assurance. Proc. IEEE **96**(6), 1000–1011 (2008)
28. Connexions: http://cnx.org
29. Rosewell, J., Ferreira, G.: QA in open educational resources (OER): open access to quality teaching resources. European Seminar on QA in e-learning, UNESCO, Paris. http://www.slideshare.net/J.P.Rosewell/qa-in-elearning-and-open-educational-resourcesoer-8398956 (2011)
30. Devedzic, V.: Education and the semantic web. Int. J. Artif. Intell. Educ. **14**, 39–65 (2004)
31. Johnson, W.L., Rickel, J., Lester, J.C.: Animated pedagogical agents: face-to-face interaction in interactive learning environments. Int. J. Artif. Intell. Educ. **11**, 47–78 (2000)
32. Hassan, S., Mihalcea, R.: Learning to identify educational materials. In: Conference in Recent Advances in Natural Language Processing, pp. 123–127, Borovets, Bulgaria(2009)
33. Hassan, S., Mihalcea, R.: Learning to identify educational materials. ACM Trans. Speech Lang. Process. **8**(2), 2–18 (2011)
34. Meyer, M., Hannappel, A., Rensing, C., Steinmetz, R.: Automatic classification of didactic functions of e-learning resources. In: 15th International Conference on Multimedia'07 (MM'07), pp. 513–516, Augsburg, Germany (2007)
35. Meder, N.: Didaktische Ontologien. Globalisierung und Wissensorganisation: Neue Aspekte für Wissen,Wissenschaft und Informationssysteme, 401–416 (2000)
36. Sanz-Rodriguez, J., Dodero Beardo, J., Sánchez-Alonso, S.: Ascertaining the relevance of open educational resources by integrating various quality indicators, RUSC—Revista de Universidad y Sociedad del Conocimiento, **8**(2), 211–224 (2011)
37. Han, K.: Quality rating of learning objects using Bayesian belief networks. PhD thesis, Simon Fraser University , Canada (2004)
38. Nesbit, J.C., Li, J.Z., Leacock, T.L.: Web-based tools for collaborative evaluation of learning resources. J. Systemics Cybern. Informatics, **3**(5) (http://www.iiisci.org/journal/sci/Contents.asp?var=&previous=ISS2829) (2005)
39. Burgos Aguilar, J.V.: Rubrics to evaluate OERs. www.temoa.info/sites/default/files/OER_Rubrics_0.pdf (2011)
40. Kumar, V., et al.: Quality rating and recommendation of learning objects. In: Pierre, S. (ed.) E-Learning Networked Environments and Architectures—A Knowledge Processing Perspective, pp. 337–373. Springer, London (2007)
41. ACHIEVE: http://www.achieve.org
42. OER Commons: http://www.oercommons.org
43. Learning Registry: http://www.learningregistry.org
44. MIT OCW: http://ocw.mit.edu/index.htm
45. OCW Consortium: http://www.ocwconsortium.org
46. The Saylor Foundation: http://www.saylor.org
47. University of Washington Courses: http://www.cs.washington.edu/education/courses
48. Coursera: https://www.coursera.org
49. Webcast.Berkeley: http://webcast.berkeley.edu
50. Universia: http://ocw.universia.net/es
51. ParisTech Libres Savoirs: http://graduateschool.paristech.fr
52. Open.Michigan: http://open.umich.edu
53. OCW University of California, Irvine: http://ocw.uci.edu/courses/index.aspx
54. OCW University of Southern Queensland: http://ocw.usq.edu.au
55. OCW Utah State University: http://ocw.usu.edu
56. Intute: http://www.intute.ac.uk/computing
57. Textbook Revolution: http://textbookrevolution.org/index.php/Main_Page
58. Google custom OER/OCW Search: http://www.google.com/cse/home?cx=000793406067725335231%3Afm2ncznoswy
59. OCW search, http://www.ocwsearch.com
60. MIT OpenCourseWare on Database Systems: http://ocw.mit.edu/courses/electrical-engineering-and-computer-science/6-830-database-systems-fall-2010

61. Saylor Foundation's Introduction to Modern Database Systems open courseware: http://www.saylor.org/courses/cs403
62. Stanford's Professor Jennifer Widom Introduction to Databases open courseware: https://www.coursera.org/course/db
63. Introduction to Database Systems courseware, Nguyen Kim Anh, in Connexions: http://cnx.org/content/m28135/latest/
64. King Fahd University's KFUPM OpenCourseWare on Database Systems: http://ocw.kfupm.edu.sa/
BrowseCourse.aspx?dname=Info.+%26+Computer+Science&did=ICS&cid=ICS324
65. University of Washington's Introduction to Data Management open courseware: http://www.cs.washington.edu/education/courses/cse344/12au
66. Universidad Charlos III de Madrid's Database Fundamentals: http://ocw.uc3m.es/ingenieria-informatica/fundamentos-de-las-bases-de-datos
67. Universidad Politecnica de Madrid's Database Administration: http://ocw.upm.es/lenguajes-y-sistemas-informaticos/administracion-de-bases-de-datos
68. Vladoiu, M.: State-of-the-art in open courseware initiatives worldwide. Informatics Educ. **10**(2), 271–294 (2011)
69. Vladoiu, M.: Open courseware initiatives—after 10 years. In: 10th International Conference Romanian Educational Network—RoEduNet, pp. 183–188. IEEE Press, Iasi (2011)
70. Brockbank, A., McGill, I.: Facilitating Reflective Learning in Higher Education. SRHE and Open University Press, Buckingham (1998)
71. Light, G., Cox, R.: Learning and Teaching in Higher Education. The Reflective Professional. Paul Chapman Publishing, London (2001)
72. Loughran, J.J.: Developing Reflective Practice. Learning About Teaching and Learning Through Modelling. Falmer Press, London (1996)
73. Schunk, D.H., Zimmerman, B.J.: Self-regulated Learning—from Teaching to Self-reflective Practice. Guilford Press, New York (1998)
74. Panait, L., Luke, S.: Cooperative multi-agent learning: the state of the art. Auton. Agents Multi-Agent Syst. **11**(3), 387–434 (2005)
75. Mitchell, T.M.: Machine Learning. McGraw-Hill Higher Education, New York (1997)
76. Sebastiani, F.: Machine learning in automated text categorization. ACM Comput. Surv. **34**(1), 1–47 (2002)
77. Rao, A.S., Georgeff, M.P.: Modeling rational agents within a BDI-architecture. In: Fikes, R., Sandewall, E. (eds.) Knowledge Representation and Reasoning (KR&R-91), pp. 473–484. Morgan Kaufmann Publishers, San Mateo (1991)
78. Rao, A.S., Georgeff, M.P.: BDI agents: from theory to practice. In: International Conference on Multi-Agent Systems (ICMAS-95), San Francisco, USA (1995)
79. Müller, J.P.: The Design of Intelligent Agents: A Layered Approach, LNCS, vol. 1177: Lecture notes in artificial intelligence. Springer, Berlin (1996)
80. Moise, G.: A software system for online learning applied in the field of computer science. Int. J. Comput. Commun. Control **II**(1), 84–93 (2007)
81. Wooldridge, M., Jennings, N.: Intelligent agents: theory and practice. Knowl. Eng. Rev. **10**(2), 115–152 (1995)
82. Sycara, K.P.: Multiagent systems. AI Mag. **19**(2), 79–92 (1998)
83. Young T.Y., Calvert, T.W.: Classification, Estimation, and Pattern Recognition. American Elsevier Pub. Co., Amsterdam (1974)
84. Archer, N.P., Wang, S.: Application of the back propagation neural network algorithm with monotonicity constraints for two-group classification problems. Decis. Sci. **24**(1), 60–75 (1993)
85. Guobin, O., Yi, LuM: Multi-class pattern classification using artificial neural networks. Pattern Recogn. **40**, 4–18 (2007)
86. McCulloch, W.S., Pitts, W.: A logical calculus of the ideas immanent in nervous activity. Bull. Math. Biophys. **5**(4), 115–133 (1943)
87. Rosenblatt, F.: The perceptron: a probabilistic model for information storage and organization in the brain. Psychol. Rev. **65**(6), 386–408 (1958)

88. Zilouchian, A.: Intelligent control systems using soft computing methodologies. CRC Press, Boca Raton (2001)
89. Moise, G., Netedu, L., Toader, F.A.: Bio-inspired E-learning systems. A simulation case: english language teaching. In: Pontes, I., Silva, A., Guelfi, A., Takeo Kofuji, S. (eds.) Methodologies, Tools and New Developments for E-Learning, p. 14. InTech, Rijeka (2012)
90. Pearl, J.: Probabilistic reasoning in intelligent systems: networks of plausible inference. Morgan Kaufmann, San Mateo (1988)
91. Bayes, T., Price, R.: An essay towards solving a problem in the doctrine of chance. Philos. Trans. R. Soc. London **53**, 370–418. www.stat.ucla.edu/history/essay.pdf (1763)
92. Friedman, N., Geiger, D., Goldszmidt, M.: Bayesian network classifiers. Mach. Learn. **29**, 131–163 (1997)
93. Baesens, B., et al.: Bayesian network classifiers for identifying the slope of the customer lifecycle of long-life customers. EJOR **156**(2), 508–523 (2004)
94. Minsky, M.: Steps toward artificial intelligence. Proc. Inst. Radio Eng. **49**(1), 8–30 (1961)
95. Piedra, N., Chicaiza, J., Tovar, E., Martinez, O.: Open educational practices and resources based on social software: UTPL experience. In: 9th IEEE International Conference on Advanced Learning Technologies—ICALT 2009, pp. 497–498, Riga (2009)

Chapter 8
E-Assessment Systems and Online Learning with Adaptive Testing

Marjan Gusev and Goce Armenski

Abstract In this paper we explain how a systematic approach to design Assessment as a Service on a cloud with SOA architecture can exploit add on functionalities like online learning with interactive assessment as a type of formative and integrative assessment for the learning process. The idea is realized by a system (computer based program) that systematically asks questions and leads the students towards knowledge construction and discovery. The system should be as simple as possible and intelligent enough to enable the realization of this idea. This online learning tool with adaptive testing uses software agents, where several strategies define the agent's behavior. We observed that the strategy "3 correct answers in a row" performs the best.

Keywords E-testing · E-learning · Computer based testing

8.1 Introduction

There are a lot of e-Learning software products, mainly realized as Learning Management Systems (LMS) that do not support e-Assessment and e-Testing with full coverage of features [14, 43]. Some significant examples include commercial products Blackboard [6] and WebCT [50], or open source software products, like Moodle [32], Fle3 Future Learning Environment [15], The Manhattan Virtual Classroom [10], LRN [30] etc.

M. Gusev (✉) · G. Armenski
Faculty of Information Sciences and Computer Engineering, Ss. Cyril and Methodious University, Rugjer Boshkovikj 16, 1000 Skopje, Macedonia
e-mail: marjan.gushev@finki.ukim.mk

G. Armenski
e-mail: goce.armenski@finki.ukim.mk

M. Ivanović and L. C. Jain (eds.), *E-Learning Paradigms and Applications,*
Studies in Computational Intelligence 528, DOI: 10.1007/978-3-642-41965-2_8,
© Springer-Verlag Berlin Heidelberg 2014

On the other hand, there are stand-alone products and designs of e-Assessment and e-Testing that have a sophisticated set of functionalities, but lack integration and interoperability with existing products. Examples include sophisticated testing centers for professional certificates, like Prometric [38] or Pearson Vue [37] and plenty of solutions that offer different types of electronic testing, questionnaires, and surveys on Internet, such as eSurveyPro [17], eSurvey Creator [16], Free Online Surveys [21], etc.

The existing LMS systems and specialized e-Testing systems are not designed to be interoperable or to act as cloud solutions and deliver Assessment as a Service, i.e. be capable to exchange tests and related data elements with other LMS systems, and therefore they can not support each other or even create new forms of efficient learning. For example, Gierlowski and Nowicki claim that vast majority of currently available electronic knowledge assessment tools are not only extremely similar but also offer strictly limited functionality [22].

In this paper we address the research problem of how e-Testing can be used efficiently in e-Learning, particularly in the construction of online adaptive learning environment and the realization of similar pedagogical methods.

Clark and Meyer define 3 essential e-Learning types as information acquisition, response strengthening and knowledge construction, using inform goals, perform-procedure goals and perform-principle goals correspondingly [9]. Aristotle and Socrates used the best method to educate the students by continuously asking them questions to stimulate critical thinking, and to illuminate ideas [5, 11]. We map these methods into knowledge construction by knowledge discovery. Our goal is to define a learning system, where the students provide answers and come to understand that they have learned the relevant knowledge by analyzing and constructing answers. The philosophy behind this idea is to select a strategy how to choose (and ask) questions and lead the students towards knowledge discovery. The final aim is to realize a computer based program which will lead the students towards knowledge discovery and construction. The system should be as simple as possible and intelligent enough to enable the realization of this idea. We call this system **online learning with adaptive testing**, while the standard e-Testing system is mainly used to assess the student knowledge as **click test**.

The online learning with adaptive testing is a system that evaluates a different test, each student every time he/she applies for. This is realized by defining a strategy for the exam and building a database with questions, which is allowed for self testing, learning, and conducting exams. The realized electronic system is based on sophisticated web technology with appropriate service oriented architecture capable to be hosted on cloud.

To realize the overall idea we also address autonomous software agents. They act as intelligent agents in the role of instructors capable to modify the way the students learn. The goal of the online learning system with adaptive testing is to set the system to act in the role of an instructor and ask questions which lead the students from one knowledge item to another in the process of knowledge construction. There are several strategies to define the agent's behavior and our research proved that the **3 in a row** strategy performs the best. The experiments realized at the end of the course showed that by using this system the students learn the relevant knowledge items faster and more easily.

The paper is organized as follows. Section 8.2 presents the state-of-the-art on e-Assessment system, its architecture and organization as cloud solution, the interoperability aspects and the knowledge database organization; then algorithms and procedures covering test question types, test delivery models, test creation algorithm and grading. Section 8.3 presents the online learning system with adaptive testing by analyzing the interactive response learning system, navigation algorithm and decision making strategy. This section also contains description of related work and coverage of software agents. The final Sect. 8.4 is devoted to conclusion and future work.

8.2 State of the Art

In this section we give overview of state-of-the-art on e-Assessment systems with correlation to the online learning with adaptive testing.

8.2.1 Computer Based Testing

Luecht and Sireci present a brief history of Computer Based Testing (CBT) [31], concluding that there is no single CBT model ideal for all educational tests, rather, all models have their strengths and weaknesses, and some are better suited to the characteristics of a particular testing program than others.

Each electronic test consists of questions that represent appropriate knowledge item (or learning objective). Seven dimensions (structure, response action, media inclusion, interactivity, complexity, fidelity, scoring method) are identified by Parshall et al. as important for e-Assessments, and a corresponding taxonomy is created [35]. Scalise and Gifford define a taxonomy of 28 innovative item types that may be useful in computer-based assessment [41]. These taxonomies address question types in e-Assessment systems.

Crisp presents a taxonomy of e-Assessment systems based on the level of constraint in the item/task response format, while analyzing the interactive assessment [12]. He observed the interaction among the teacher (instructor) and the student, and defined its goal to be "assessment **for learning**" or "assessment **of learning**". Sclater and Howie identify credit bearing tests, self assessment and diagnostic tests according to their purpose and possible use in interactive assessment [42].

Table 8.1 presents the correlation of our definitions to these categorizations.

There has been intensive research and development in the field of e-Learning systems. For example, Davies and Davis in [14] present the results of EU funded projects in using grid infrastructures for e-Learning and technologies for online

Table 8.1 Correlation of different taxonomies to online learning with adaptive testing and click tests

Source	Online learning with adaptive testing	Click test
Purpose in our definition	Adaptive learning	Assessment
Category by Sclater and Howie [42]	Self assessment and diagnostic testing	Credit bearing testing
Category by Crisp [12]	Formative assessment	Diagnostic and summative tests

interoperable assessment. Dagger et al. define various characteristics of modern LMS using adaptive hypermedia and semantic web technologies based on service oriented architecture [13].

8.2.2 Architecture and Design

In this section we give an overview of a modern e-Assessment system, its architecture consisting of a set of services used in an e-Learning context and collectively realize required business objective. We also address interoperability as a very important feature of this system.

8.2.2.1 Modern E-Assessment System Architecture

There are a lot of papers and projects describing modular e-Assessment architecture. For example, Gierlowsky designed a highly scalable and modular architecture with several layers and modules [22]. Armenski and Gusev designed a three-layered architecture to capture most of the demands of a modern e-Assessment system [2] as presented in Fig. 8.1. This architecture is intended to be used by any

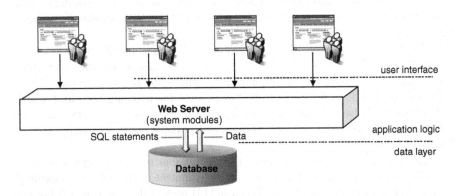

Fig. 8.1 Three layered architecture of the system [2]

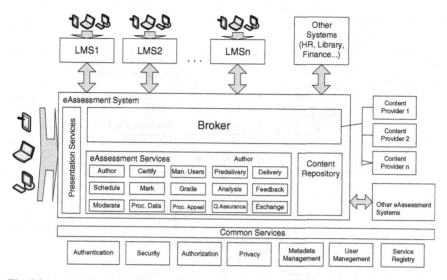

Fig. 8.2 Architecture of an ultimate e-assessment system [3]

computer on Internet through common web browser user interface, and separates the database layer from the application layer for basic system modules and also from the user interface layer.

Furthermore, they have presented a novel architecture based on SOA to support all relevant functions in computer based assessment [1] and defined an ultimate e-Assessment system with more advanced approach trying to define the ultimate assessment engine, the architectural style behind, and its overall architecture [3], as presented in Fig. 8.2. Although this model allows construction of a stand-alone system, it offers a possibility to construct a Software as a Service cloud solution, by defining a broker module that establishes communication to different LMS systems using various standards and acts the role of system service orchestrator.

Recently, Ristov et al. defined a sophisticated cloud solution on top of these ideas [39], by defining three subsystems, i.e. the Management, Assessment and Reporting subsystems, as presented in Fig. 8.3 and an organization-scheduling algorithm for different Virtual Machines (VMs), instead of one broker. The Admin, Student and Reporting agents serve to enable communication among different virtual machines, and communication between agent and administrator, teacher and student correspondingly. A special agent module, called Infrastructure agent is responsible for resource provision, i.e. activates or cuts provision of a VM by analyzing the workload.

8.2.2.2 Interoperability

Regarding the interoperability of e-Learning systems, Dagger et al. differentiate three generations [13]. However, interoperability has not been addressed thoroughly, although several detailed frameworks support standardization efforts in the

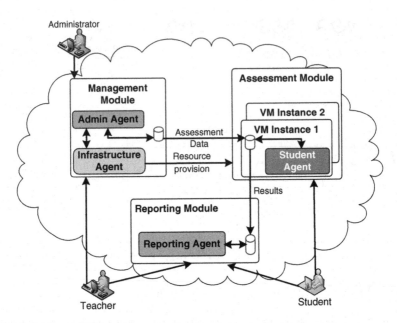

Fig. 8.3 E-assessment system organization [39]

e-Learning domain, like JISC e-Learning Technical Framework (ELF) [19], IMS Abstract Framework (IAF) [24], Open Knowledge Initiative (OKI) [33], LeAPP Learning Architecture Project [29] etc. Another example is the project [45] with goal to establish appropriate infrastructure to develop competences, addressing an assessment model, without supporting services [44].

Although service oriented architecture is mostly implemented trend in the last years, still there are no commercial e-Assessment systems on the market following the real system architecture based on SOA, which is ready for interoperability challenges for the growing demands in the world. Mainly this is due to the lack of standards and a lot of industry pushed solutions which in essence do not like to be interoperable.

In the beginning, the organizations formed consortiums to define interoperable standards, and while all industry players were implementing them, there was an on-going debate about cloud computing and a possibility for interoperable services on top of these solutions. Therefore still interoperability of e-Education systems is a hot research topic with a lot of research on standards and work on the development of new solutions is underway. The results of all these efforts will definitely have impact and influence on all new software solutions.

The capability of a given software system to use the same formats to store and retrieve information means it uses **interoperable** formats. In addition **interoperability** of services means that two different software systems realized on various hardware and platforms can interchange information or services [24]. Dagger et al. [13] discusses two levels of interoperability between LMS and its tools: interoperability of content and interoperability of tools. Vossen and Westercamp [49] identified one more level of interoperability in exchanging user data. Several published standards like SCORM [28], IMS Content Packaging [25] and IMS Learning Design [26] are evidence of intensive research about content interoperability recently.

The presented model of e-Assessment system [3] follows the trend to separate the content from tools, making possible a seamless and dynamic exchange of tools, functionalities, semantics and control. The specified system is built with Service Oriented Architecture, based on encapsulating existing business functions as loosely coupled, reusable, platform-independent services which collectively realize the required business objective. The final goal is to enable a system that uses widely adopted standards, increases system flexibility and supports a lot of pedagogical diversities.

Another innovation that the model in [3] puts an accent on is the concept of pluggability. In a real system, this means that the interoperability is already established and the system architecture can be pluggable to other systems both in educational and business environments. An e-Assessment system can be pluggable if it has ability to be easily attached to any existing LMS and its functionalities to be used as if that e-Assessment system was part of the LMS from its basic installation.

Cloud computing offered a completely new perspective on the development of solutions. For example, if an e-Assessment system is built such that it is pluggable to any LMS, it should not only offer a complete service, but it should also offer subsets and various sub services, like realization of questionnaires, inquiries, public opinion gathering service etc. To make this possible, the system should use a highly interoperable service oriented architecture, building modular services that can exchange interoperable information. For example, a company that uses a structured set of customers and would need to gather customers' opinion through a questionnaire. The company probably would like to exchange the set of customers within the accounting software service, send e-mails via e-mail marketing service and use e-Testing software service for questionnaires. All these services should be interoperable and exchange required information.

A modern LMS should support the exchange of learning resources and profiles with other systems including the legacy systems. It will also support an extensive usage of e-Testing for learning process, rather than just for assessment. The exchange includes knowledge items, hypermedia content and personalized learning environments. Finally, it will benefit with increased efficiency and effectiveness. For example, Thurlow et al. give benefit overview [48].

8.2.3 Algorithms and Procedures

This section gives an overview of used algorithms and procedures for the e-Testing process in an e-Assessment system. It includes organization of knowledge items and questions, test delivery methods and test creation algorithms.

The knowledge items are stored in a knowledge database, organized usually as a tree, where each lecture consists of parts, each part, of sets, each set of learning objectives, etc. The final leafs are questions that correspond to a given knowledge item. In addition to basic information provided by the question, an extensive information can be stored for each question in the knowledge database including statistics of realized testings and answers given.

Tests are created according to test creation algorithms, which form tests by selecting knowledge items from the knowledge database and then the e-Testing software continues to realize the testing by presentation of questions and collecting the student answers via test delivery models. Patelis gives an overview of various test delivery models [36] identifying linear tests, dynamic linear, testlets, mastery models, and adaptive tests. Thompson and Wiess identify three primary variable-form approaches: computerized adaptive testing (CAT) and linear-on-the-fly testing (LOFT), and multistage testing [47]. Luecht and Sirreci differ eight CBT models primarily with respect to their use of adaptive algorithms, the size of the test administration units, and the nature and extent to which automated test assembly is used [31].

8.2.3.1 Test Creation Algorithm

Test creation algorithm is closely connected to the chosen method for test delivery. The idea for creating different tests for every student, forced us to apply the model for dynamic test creation. With that idea every student will get different test, with same weight like all other students. These dynamically created tests will have a fixed number of questions because this was first time system for automated assessment to be applied at our University. In order to provide a less painful change in the way of taking the assessment and to lower the difficulties in its adaptation we have decided that fixed number of question is a better solution than dynamic one. The same reason forced us to use dynamic test creation model instead of model for adaptive testing because of the easiness and the transparency that non adaptable test have. The applied model gives an opportunity for students to list the questions one by one, and answer only those whose answer they know.

The strategy for test generation is defined by the course administrator when he schedules the assessment. The course tree structure is used to set a strategy. The administrator is marking the learning objects from which questions will be selected, specifying the number of questions taken from each learning object. So, the course administrator will have control over the curricula. Since each learning object has questions with same weight, the tests which will be generated will have same weight too, but the students will get different tests from those learning objects selected by the course administrator. The system has a feature with which already made strategies are saved and can be used in the future.

Test creation algorithm may be rather complex for adaptive testing and variable-form testing with algorithmic approach, where the test is designed to be administered with a dynamic, interactive algorithm, in contrast to multistage testing, where there are fixed routes between the testlets [47]. For example, AS method uses an adaptive algorithm that maximizes the test information function for each examinee. This approach leads to overexposure of a relatively small portion of the entire test bank since the most informative items are continually in high demand. AT algorithm is based on the idea that after completing the testlet, the computer scores the items within it and then chooses the next testlet to be administered. Thus, this type of test is adaptive at the testlet level rather than at the item level.

8.2.3.2 Grading

Automated scores are consistent with the scores from expert human graders, they are fair and have been validated against external measures in the same way as is done with human scoring, according to [51].

Grading as a process starts when the testing is finished, i.e. when the student submits the final answers to the system or when the time limit is exceeded. It is a process of evaluation of the submitted answers by matching the answers against correct answers stored in the database.

The evaluation process in case of fixed response questions is realized as a straight forward procedure of checking if the answer is correct. There is only one constraint since the test allows rotation of possible answers, so the check is performed against real correct answers, instead of a given schema.

A special procedure starts to evaluate the answer in the case of answers where additional computation is required, or in case of essay questions. Often these procedures are realized with human interventions and decisions.

In our system, final results are stored in a database and sent to the display system. The system displays the final results in two ways, either using points or percents. There is an option to see the right answers compared to those entered by the user. In such a way the student realizes where the mistake is and finds the correct answer. This is a method of assisted learning, usually used by teachers with corrections whenever a mistake is noticed.

A lot of efforts were made to enable security of the system. One problem that arises often is the method of guessing. The students do not try to answer a given question, instead, they are trying to guess the answer by random clicking on a possible answer. In a long term education process, as a kind of repetition method, the student will, hopefully, learn the corresponding knowledge item. However, in order to eliminate guessing, we have implemented negative marking. This is a technique where each mistake is punished by achieving negative grade. Applying this procedure the students now avoid guessing and answer the questions only if they are certain about the correct answer.

8.3 Description of the Online Learning with Adaptive Testing

Here we will explain the details of the online learning system realized with adaptive testing. In this system, there is a high interaction between the teacher and students, integrating the e-Testing solution into an efficient learning process. The basic idea is explained in Sect. 8.3.1 as interactive response learning system, while the navigation through the lecture and knowledge items in Sect. 8.3.2 and the decision strategy algorithm in Sect. 8.3.3.

8.3.1 Interactive Response Learning System

E-education refers to activities connected to the process of increasing the students's knowledge and skills. E-assessment is not just referring to the process of evaluation of students knowledge, but it can be efficiently used in e-Education. New services to be developed in a ultimate modern e-Assessment system should also have ability to integrate and embed in the learning context or in the process when the student increases the knowledge. It usually requires a higher level of student involvement in the assessment and learning process. It leads to an idea about a complex learning process where teaching, learning and assessment interact frequently to increase the efficiency and overall impact. To realize this new learning approach, the e-Assessment services should be highly integrated in e-Education processes and be capable to address the complex challenges and characteristics. The system proposed in [23] is identified as interactive response learning system or just online learning tool.

The basic idea is based on a scenario presented in Fig. 8.4. A student chooses to attend a course lectures online, makes a subscription and creates a record for his activities. Then, the student can navigate the course tree structure and starts with the online learning tool. Learning materials are presented for each lecture as part of the LMS and than test generation starts to evaluate the students's knowledge about particular learning objectives, knowledge items or skills.

This scenario is typical for each student. In this case, the impact of the lecturer is built in the system. The lecturer usually provides a corrective measure, i.e. corrects the students if there is a mistake or approves correct answers. This is exactly what the online learning tool with adaptive testing does. However, there is a strategy when the teacher will continue to the next learning objective or explanation of a new knowledge item. Usual behavior of a good teacher is to move to the next level only if the student has learned the previous learning objectives and knows all relevant knowledge items in order to be capable to follow the next item. This has to be done in a way that the student does not loose concentration or gets no frustration if not all questions are answered correctly, but has sufficient understanding about the corresponding learning objective. This is why we have to

Fig. 8.4 Online learning
scenario with adaptive testing

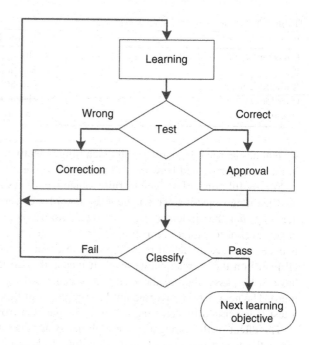

implement adaptive testing which adapts to the current knowledge of the student, a situation that will simulate the teachers way of thinking.

The realization of this scenario uses several algorithms for adaptive testing and usage of software agents. The navigation through the learning objectives should follow a predefined path, but also be capable to enable different scenarios to reach the final goal, for example, better students with good background would like to choose a shorter path and avoid variations or similar questions, and the students that are not motivated and with no background should probably follow the longer and exhaustive path. All these algorithms will be discussed in the next sections.

8.3.2 Navigation Algorithm

The navigation through the learning objectives is possible for the already explained knowledge database with tree-like structure. Lectures are branches and knowledge items are leafs in the tree. Each lecture consists of smaller parts and each part consists of different sets and finally of learning objectives. The constraints used in this tree-like knowledge database structure are presented in Table 8.2.

The navigation can start in the leftmost nodes in the tree. The questions are selected out of possible candidates in the given set/learning objective. The testing process is realized by asking questions one by one in sequential order. There is a strategy which decides whether the provided answers are sufficient to show that the

Table 8.2 Constraints in the knowledge database

Item	Constraint
Course material	Has at least 3 parts
Part	Has at least 3 knowledge items
Knowledge item	Has at least 4 question sets
Question set	Has at least 5 questions
Question	At least one is hidden for final exam

student has learned the corresponding learning objective. This strategy is explained in the next section and here we will present how the navigation can continue.

Successful finish of a given learning objective allows the student to move to the next learning objective. This is done by preorder traversal in the tree i.e. L-R-O, meaning that the student finishes the leafs from left to right and then moves to upper level or to another branch in the tree-like structure. Moving to the next level is once more accompanied by a summary test that selects at least one question of all nodes on the corresponding level. It means that all the sets from a given part have to be passed before the final test is generated for a given part (knowledge item). The next part is traversed until a complete fulfillment of lecture. Once more, a test is generated by selecting at least one question from each previous node.

The effect of this navigation through the system is interesting for the students. Since the test generation algorithm randomly chooses a question from a given knowledge item and randomly presents possible answers, the students have a feeling that the system is different for all of them. In a sense, this also happens with a real teacher. Another issue is the navigation path, since there are more than 3 nodes on same level, the movement to next node is also random, so each student, in effect, has different navigation path, i.e. learning path.

Once the student passes all knowledge items on the same level, the system sets a test for the given part, now randomly selecting a question from each node. Here the system might ask the very same question, which is allowed following the strategy that repeating enables better learning. Although the first impression of the students is that the system asks the same questions and they appear randomly, in the background, the system is navigating through the knowledge database tree and acts as a teacher who is always asking questions, corrects the wrong ones and awards the correct answers by letting the students to upper levels.

All navigation paths are stored in the system and also all statistics about given answers. The lecturer knows which students are subscribed and can view their records and analyze their achievements.

8.3.3 Decision Strategy

A decision making strategy to evaluate if the student has learned a certain learning objective or knowledge item is very important. There are several issues that have to be analyzed before developing such a strategy.

If one question is set from a given knowledge item (we assume there are at least 3 questions for a given learning objective), we may expect guessing as a method of answering. In this case we are not sure if the student really learned the corresponding knowledge item or not.

If all questions are asked, then we will be sure that the student has answered all of them, but this leads to exhaustive navigation through all questions. In this case, especially the better students will be bored. So the only alternative is to find an appropriate strategy that will decide when the student has learned the corresponding knowledge item.

Another issue is raised for a situation when a wrong answer is given by the student. The system gives the result after each answer. This works as a corrective measure providing valuable feedback, so the student will now know the correct answer to a given question. However, this does not mean that the student has learned the corresponding knowledge item, just that the correct answer was shown to the student and the method of recognition might be used instead of presentation of real knowledge.

The only alternative to realize this adaptive strategy is based on correctly answered questions and their order. For this purpose we have set an experiment and tested which strategy the students think is the best in the learning process. The experiment was to test the decision strategies presented in Table 8.3.

The decision strategy enables the online learning system with adaptive testing to decide when to jump from one to another learning objective. The strategies AC and 1C are easily understood, since they refer to correctly answering of all or one question. The strategies 3C and 3R assume that the student should answer three questions correctly. The difference is in how these correct answers are obtained. 3C just counts correctly answered questions in a given learning objective and makes positive decision without checking if the student has already answered wrongly some questions, while 3R counts in-a-row, so if three questions are answered correctly consequently (in-a-row) then the decision is made. For example, let the learning objective have 10 questions and the obtained answers are wrong, two correct, wrong and three correct or WCCWCCC, where W means wrong and C correct answer, then the 1C strategy makes decision after the second question, 3C after the fifth question and 3R after the seventh question. Our experiments showed that 1C and 3C strategies are vulnerable to guessing, while 3R is less.

Table 8.3 Decision strategies to move into another learning objective

ID	Strategy	Description
AC	All correct answers	The student should answer correctly all questions for a given learning objective
NC	N correct answers	The student should answer correctly N questions for a learning objective, although some questions might be not answered correctly
NR	N correct in-a-row	The student should answer correctly N questions for a learning objective consequently (one by one in-a-row)

The best alternative 3R strategy is realized by the following algorithm. If the student answer is not correct, the counter for correct answers in a row is reset. Then the questions are repeatedly selected randomly within a given knowledge item. The decision is made only after three consecutive correct answers.

This adaptive strategy concerns not only correct answers in a row, but also timing constraints and level of difficulty. After easy questions are set, the difficult ones follow in an adaptive manner according to series of correct answers.

In our system we implemented the adaptive strategy, which records the student achievement. If the student is knowledgable and answers correctly in three consecutive parts, then the strategy is adapted to "2 correct answers in a row" and in that sense it moves to "one correct answer in a row" meaning one question per learning objective. We believe that this strategy is suited to better students and will eliminate the syndrome that the testing process is boring. It will still be motivating for the students.

There is a record of the efficiency of the tests the student takes and of the navigation paths. This data can be presented to the teacher and also to the student.

The goal of e-Assessment should be specified explicitly. If the final aim is to assess student knowledge and skills then we should use a strategy that classifies the grades according to predefined criteria, as it is done by ETS [18]. However, the goal of the online learning system with adaptive testing is to provide a proof that the student has passed over all learning objectives showing correct answers for relevant knowledge items. It does not grade or classify the grades, nor evaluates the student efficiency. It is intended as a support tool for the students to learn the corresponding learning objectives.

8.3.4 Software Agents

There are a lot of definitions about software agents as computer programs that act for a user with authority to decide. Russel and Norvig define intelligent agents as an abstract functional system similar to a computer program. According to their definition [40], intelligent agent is an autonomous entity which observes through sensors and acts upon an environment using actuators and directs its activity towards achieving goals (i.e. it is rational), as presented in Fig. 8.5. The online learning tool with adaptive testing can be classified as intelligent agent since the agent (online learning tool with adaptive testing) takes action of the environment (student's knowledge) by perception (assessment) of current student's knowledge and skills. The sensors (e-Testing system) send information to the core where the actions and decisions are made (whether the student has learned the corresponding learning objective) and actuators start actions (the student can move to next level—learning objective).

Another definition in [20] defines that an autonomous agent is a system situated within and a part of an environment that senses that environment and acts on it, over time, in pursuit of its own agenda and so as to effect what it senses in the future.

Fig. 8.5 Simple reflex
intelligent agent [40]

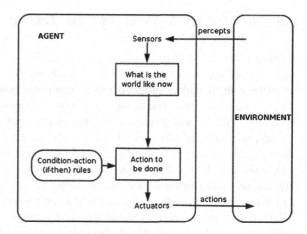

Table 8.4 Autonomous software agent properties that our system has

Property	Description
Reactive	Responds to learned knowledge items
Autonomous	Controls the process of learning (moving from one learning objective to another)
Continuous	Runs as continuous process over time
Communicative	Interacts with students
Adaptive	Changes its behavior based on current state

According to the definition given in [20], our system acts as an autonomous software agent, since the environment is the students knowledge, sensing correlates to assessment, and action that produces state in environment is learning. Drives are built-in preferences and act as primitive motivators. The predefined goal in the online learning tool with adaptive testing is to have approval that the student has visited all learning objectives and has proved to have initial knowledge about the knowledge items. The goal is not to have a precise assessment and classify the student's knowledge and skills according to a given schema, as it is in the regular assessment software packages. The agent's action-selection mechanism is based on the described decision strategy and navigation algorithm, allowing the student to move to the next level of learning objectives whenever the predefined criteria is satisfied.

The described online learning system with adaptive testing has several agent's properties as presented in Table 8.4.

A good overview of definitions about software agents, their classification and architectures is given by Bădică et al. [4]. The authors give a comprehensive list of agent features, although in Table 8.4 we refer only to specific properties analyzed by our system.

8.3.5 Related Work About Adaptive Testing

Clark and Meyer define asynchronous e-Learning with feature to dynamically tailor instruction to the changing needs of learners. This feature is realized by an adaptive control with a program that dynamically adjusts lesson difficulty and support based on the evaluation of learner's response [9]. They provide an extensive evidence and references for dynamic adaptive control.

Adaptive learning using web has been analyzed a lot in the literature. Brusilovsky gives an overview of adaptive and intelligent technologies for web-based education and identify five major technologies with immediate roots in Adaptive Hypermedia and Intelligent Tutoring Systems [7, 8]. According to their classification our definition of the online learning system with adaptive testing belongs to adaptive navigation support (assisting in hyperspace orientation), curriculum sequencing (finding optimal path through the learning material) and intelligent solution analysis (using intelligent classifiers with extensive error feedback).

Paramythis and Loidl-Reisinger give overview of Adaptive Learning Environments (ALE) and e-Learning standards [34]. They define 4 categories of adaptation in learning environments and our definition of online learning with adaptive testing belongs to content discovery and assembly, where the adaptive component lies in monitoring of student's knowledge.

In this paper we address adaptive testing, where a specific algorithm will adapt the test to the knowledge and skills of each student, not just with respect to selected items by the tester (teacher). According to [46] the basic CAT method is an iterative algorithm that searches for the optimal question, based on the current estimate of the student's score, and after answering correctly or incorrectly, the score is updated.

So, the method is to choose the optimal question based on current score, and dynamically to update the score. There are different strategies to choose the optimal question, but in our model of online learning with adaptive testing we refer to two step procedure, first we would like to be sure that the student has learned the corresponding learning objective by not asking all questions, and navigating the knowledge tree as method where to move next.

The final goal in adaptive testing is also very important. Most of the existing strategies for adaptive testing aim to classify the student's answers and make precise evaluation of his/her knowledge. Adaptive testing in this case should always adapt to the knowledge level the student shows and the system should ask questions from upper and lower levels in order to make a final decision about the level of knowledge. For example, up-down methods are used in most of the existing testing software products [18] and there are a lot of variants, such as, 1 up 2 down, 1 up 3 down, etc. Kaernbach introduces simple adaptive testing with the weighted up-down method, where each correct answer leads to the higher level and each wrong answer leads to lower level [27]. There are a lot of examples of implementation of this strategy and the final score is obtained by weighted sum of number of questions answered in appropriate level.

However, our online learning tool with adaptive testing aims to a different goal. The previous strategy corresponds to upgrade of 1C strategy which now will return backwards one knowledge item if wrong answer is obtained. Since the goal is not the evaluation of score and grading the student, but verifying if corresponding learning objective is learned by a student, then this strategy will not give better results than those we have experimentally shown.

Online learning with adaptive testing acts as software agent that represents the teacher who sets the lecture and approves all leaves and branches in the tree structure for the given lecture. This schedules the possible traversal path and lists all learning objectives that the student is expected to learn within the given lecture.

CAT provides an accurate point estimation of individual ability or achievement. Another approach is computerized classification testing (CCT) usually used to classify students according to their knowledge. Typical examples include pass/fail or basic/proficient/advanced levels. In online learning model with adaptive testing we use only pass/fail assessment by applying the decision strategy.

A properly designed and implemented CAT can affect the motivation of students [47]. Because lower-ability students will receive easier items, they will become less discouraged and stressed. Conversely, high-ability students will not be wasting time on items that are far too easy; they will receive items that appropriately challenge their high ability. This is also built in our online learning with adaptive testing which gracefully improves the strategy for better students.

8.4 Conclusion and Future Work

Initially, the design of the e-Testing system started in 1999 to help the realization of frequent assessments (each month) where more than 500 students take part. The original idea to create a system that can help the realization of exams for large number of students was expanded to realize an independent system for electronic testing. New features added values to realize not just an assessment tool, but also to support the learning process and overall education. Later on, the development of software as a service was a great challenge, especially in solving the interoperability issues.

The presented e-Testing system is implemented and in use from 2001 at the University Ss Cyril and Methodius. Added-values, such as, online learning tool with adaptive testing and self testing were installed in 2002, and a new version was released in 2003. A more sophisticated newer version was installed in 2006, completely changed using the service oriented architecture and in 2009 the system was upgraded to the cloud computing model. In the process of developing the assessment system and introducing the online learning system with adaptive testing we have faced and successfully solved a lot of problems, especially those that arise due to cheating. We have tested several strategies arriving at the best. The design of an interoperable cloud solution will enable this tool to be used as add-on in the existing LMS.

During the last 10 years 9,132 students and 110 teachers—professors from the University were registered to use the system. The database consists of 74 courses with 27,027 questions and 107,116 offered answers, i.e. average 4 answers per question. 63,255 tests were processed and 1.585.126 questions were set to students, i.e. each exam consists of average 25 questions. 51 courses were using the online learning tool with adaptive testing. On average each of these online courses has 81 parts and 900 learning objectives—knowledge items. Each student has answered on average 4, 5 questions per learning objective (knowledge item) using the 3 in-a-row strategy. Once the students passed this type of online learning with adaptive testing they successfully passed the exam without a lot of effort. It really helped them to fulfill their knowledge.

The system is interactive with course material since if a wrong answer is obtained the student goes back to the material to learn the concepts and comes back to the system. The process where students realized the self-testing led to more motivation, since they tested their knowledge more often and this accomplished the learning process more successfully.

In this paper we presented state of the art about e-Assessment systems. We have analyzed details of organization of e-Testing, and compared our definition to other e-Assessment systems addressing the interactive assessment. We also address the architecture and organization of a cloud solution, and coverage of interoperability standards and recommendations, aiming that the online learning system with interactive testing can be used as a tool or upgrade to the existing LMS systems.

The online learning system with adaptive testing was presented with specification of main requirements organization of the knowledge database, algorithms and procedures for testing process, including test delivery models, test creation algorithm and grading strategy. A special section is dedicated to the navigation algorithm and the decision making strategy. The solution uses software agent technology and classification algorithm for navigation through learning objectives and realizes appropriate decision making strategy.

The experiments were analyzed to conclude when the student has learned corresponding learning objective. The comparisons showed that 3R (**3 correct answers in a row**) is the best both for the students and the teacher. The conclusions are brought by analyzing the interviews with students and teachers about their motivation and impact. Several conclusions also reflect their attitude that the online learning system with adaptive testing actually implements:

- Students like fun and entertainment, i.e. the system is a kind of a computer game in the learning process.
- Students like challenge, freedom, unexpected elements, they like competition and this system offers it.
- Any strategy trying to repeat the same question until a correct answer is given makes the system a boring place.

This paper summarizes how e-Assessment systems are built, and how e-Testing can be implemented for e-Education, putting the accent on new trends in technology and implementation of AI related methods to establish better learning system. The final goal is to realize how e-Testing supports the e-Education making the "e" in e-Education to stand for **efficient** instead of just electronic.

In near future we plan to make advanced research on the discussed strategies, including more statistical tools and AI related methods. We have started to develop a new cloud based solution in 2012 and the process is ongoing, trying to be consistent with existing data and system. The application of the online learning system with adaptive testing does not depend on whether the system is cloud based or not, although the cloud based solution will enable to use it as a tool in different environments and LMS, by exchanging appropriate information.

References

1. Armenski, G., Gusev, M.: E-testing based on service oriented architecture. In: Proceedings of the 10th CAA International Computer Assisted Assessment Conference, pp. 17–23. © Loughborough University (2006)
2. Armenski, G., Gušev, M.: Infrastructure for e-testing. Facta Univ. Ser. Electron. Energ. **18**(2), 181–204 (2005). http://factaee.elfak.ni.ac.yu/fu2k52/contents.html
3. Armenski, G., Gusev, M.: The architecture of an ultimate' e-assessment system. In: ICT Innovations 2012, Web Proceedings. Association for Information and Communication Technologies ICT-ACT (2009)
4. Bădică, C., Budimac, Z., Burkhard, H.D., Ivanović, M.: Software agents: Languages, tools, platforms. Comput. Sci. Inf. Syst./ComSIS **8**(2), 255–298 (2011)
5. Beck, J., Stern, M., Haugsjaa, E.: Applications of ai in education. Crossroads **3**(1), 11–15 (1996)
6. Blackboard: (2013). http://www.blackboard.com
7. Brusilovsky, P.: Developing adaptive educational hypermedia systems: From design models to authoring tools. Authoring tools for advanced technology learning environment, pp. 377–409. Kluwer Academic Publishers, Dordrecht (2003)
8. Brusilovsky, P., Peylo, C.: Adaptive and intelligent web-based educational systems. Int. J. Artif. Intell. Educ. **13**(2), 159–172 (2003)
9. Clark, R., Mayer, R.: E-learning and the science of instruction: proven guidelines for consumers and designers of multimedia learning. Pfeiffer (2011)
10. Classroom, T.M.V.: (2012). http://manhattan.sourceforge.net
11. Collins, A.M., Stevens, A.L.: A Cognitive Theory of Interactive Teaching. Bolt Beranek and Newman Inc, Cambridge (1981)
12. Crisp, G.: Interactive e-assessment: Moving beyond multiple-choice questions. Centre for Learning and Professional Development. Adelaide (Australien): University of Adelaide (2009). http://ipac.kacst.edu.sa/eDoc/2009/173798_1.pdf [10.11.2010]
13. Dagger, D., O'Connor, A., Lawless, S., Walsh, E., Wade, V.: Service-oriented e-learning platforms: from monolithic systems to flexible services. IEEE Internet Comput. **11**(3), 28–35 (2007)
14. Davies, W.M., Davis, H.C.: Designing assessment tools in a service oriented architecture. In: 1st International ELeGI Conference on Advanced Technology for Enhanced Learning, Vico Equense-Naples, Italy, 15–16 Mar 2005, pp. 1–7 (2005)
15. Environment, F.L.: (2012). http://fle3.uiah.fi
16. eSurvey Creator: (2013). http://www.esurveycreator.com
17. eSurvey Pro: (2013). http://www.esurveypro.com
18. ETS: (2012). https://www.ets.org/
19. EL Framework: (2012). http://www.elframework.org/

20. Franklin, S., Graesser, A.: Is it an agent, or just a program? A taxonomy for autonomous agents. Intelligent Agents III Agent Theories, Architectures, and Languages, pp. 21–35 (1997)
21. Free Online Surveys: (2013). http://www.freeonlinesurveys.com
22. Gierłowski, K., Nowicki, K.: A highly scalable, modular architecture for computer aided assessment e-learning systems. Distance Education Environments and Emerging Software Systems: New Technologies, IGI Global (2011)
23. Gusev, M., Armenski, G.: On-line learning and etesting. In: Information Proceedings of the 24th International Conference on Technology Interfaces ITI 2002, pp. 147–152. IEEE (2002)
24. IMS Global Learning Consortium: Abstract Framework: (2012). http://www.imsglobal.org/af/index.html
25. IMS Content Packing Specification: (2012). http://www.imsglobal.org/content/packaging/
26. IMS Learning Design Specification: (2012). http://www.imsglobal.org/learningdesign/
27. Kaernbach, C.: Simple adaptive testing with the weighted up-down method. Atten. Percept. Psychophys. **49**(3), 227–229 (1991)
28. Koper, R., van Es, R.: Modelling units of learning from a pedagogical perspective. Online Educ. Using Learn. Objects. **40**, 43−58 (2004)
29. LeAP Project Case Study: Implementing Web Services in an Education Environment: (2012). http://www.education.tas.gov.au/admin/ict/projects/imsdoecasestudy/LeAPProjectCaseSummary.pdf
30. LRN: (2012). http://dotlrn.org
31. Luecht, R., Sireci, S.: A review of models for computer-based testing (2012). http://research.collegeboard.org/publications/content/2012/05/review-models-computer-based-testing
32. Moodle: (2013). http://moodle.org
33. OKI Project: (2012). http://www.okiproject.org
34. Paramythis, A., Loidl-Reisinger, S.: Adaptive learning environments and e-learning standards. In: Second European Conference on E-Learning, pp. 369–379 (2003)
35. Parshall, C., Davey, T., Pashley, P.: Innovative items for computerized testing. In: van der Linden, W.J., Glas, C.A.W. (eds.) Elements of Adaptive Testing, Statistics, for Social and Behavioral Sciences, pp. 215–230. Springer, Berlin (2002)
36. Patelis, T.: An overview of computer-based testing. The College Board, RN-09, Office of Research and Development (2000). http://www.collegeboard.com/research/html/rn09.pdf
37. Pearson Vue: (2013). http://www.pearsonvue.com
38. Prometric: (2013). http://www.prometric.com
39. Ristov, S., Gusev, M., Armenski, G., Bozinoski, K., Velkoski, G.: Architecture and organization of e-assessment cloud solution. In: 2013 IEEE Global Engineering Education Conference (EDUCON), pp. 736–743 (2013)
40. Russell, S., Norvig, P.: Artificial Intelligence: A Modern Approach, 3rd edn. Prentice Hall, Englewood Cliffs (2002)
41. Scalise, K., Gifford, B.: Computer-based assessment in e-learning: A framework for constructing "intermediate constraint" questions and tasks for technology platforms. J. Technol. Learn. Assess. **4**(6), 5−44 (2006)
42. Sclater, N., Howie, K.: User requirements of the ultimate online assessment engine. Comput. Educ. **40**(3), 285–306 (2003)
43. Sclater, N., Low, B., Barr, N.: Interoperability with CAA: does it work in practice? In: Proceedings of the Sixth International Computer Assisted Assessment Conference, Loughborough University, 317−326 (2002)
44. Tattersall, C., Hermans, H.: Ounls assessment model. In: January 2006 TENCompetence WP6 Meeting (2006). http://dspace.ou.nl/handle/1820/558
45. TENCompetence—Building the European Network for Lifelong Competence Development: (2012). http://www.tencompetence.org/
46. Thissen, D., Mislevy, R.: Testing algorithms. In: Wainer, H., Dorans, N., Green, B., Steinberg, L., Flaugher, R., Mislevy, R., Thissen, D. (eds.) Computerized Adaptive Testing: A Primer. Lawrence Erlbaum Associates Inc, London, 101−133 (2000)

47. Thompson, N., Wiess, D.: Computerised and adaptive testing in educational assessment. Trans. Comput. Based Assess. New Approach. Skills Assess Implications Large-Scale Test. 127–133 (2009)
48. Thurlow, M., Lazarus, S., Albus, D., Hodgson, J.: Computer-based testing: Practices and considerations. Synth Rep. **78** (2010)
49. Vossen, G., Westerkamp, P.: E-learning as a web service. In: Proceedings of the Seventh International Database Engineering and Applications Symposium, pp. 242–249. IEEE (2003)
50. WebCT: (2013). http://www.webct.com
51. Williamson, D.M., Bennett, R.E., Lazer, S., Bernstein, J., Foltz, P.W., Landauer, T.K., Walter, D., Way, D., Sweeney, K.: Automated Scoring for the Assessment of Common Core Standards (2010). https://www.ets.org/s/commonassessments/pdf/AutomatedScoringAssess CommonCoreStandards.pdf

Chapter 9
Mechanism for Adaptation of Group Decision-making in Multi-agent E-Learning Environment

Denis Mušić

Abstract Intense and stressful group decision-making has become a daily activity in the modern business environments which caused greater interest in systems that allow simulation of group decision-making with agents as human representatives (surrogates). Development of representative agents is significantly enhanced through use of methods that allow mapping of some of the most important human traits in the world of agents. These traits are emotions, personality and mood which gain importance by their direct effect on the process of individual and therefore group decision-making. In order to provide more stable and efficient group decision-making, this chapter presents the research results of applying concepts of experience and patience to the emotional agents in eLearning environment. Concept of experience is implemented by using Reinforcement learning technique called Q-learning in combination with Self-organizing map, while concept of patience is implemented by introducing a Self-regulation coefficient.

Keywords Agents · Patience · Q-learning · SOM · Self-regulation coefficient

9.1 Introduction

With the advent of the first movie achievements, the term agent has brought significant amount of mystery, and represented a synonym for an undercover individual that performs various kinds of tasks showing remarkable degree of intelligence and ability. With the primary objective to create more realistic human representatives, intelligence has become the standard trait in the area of software agents which allows them to perform different types of services such as: finding

D. Mušić (✉)
Faculty of Information Technologies, University Dzemal Bijedic, Mostar,
Bosnia and Herzegovina
e-mail: denis@fit.ba

M. Ivanović and L. C. Jain (eds.), *E-Learning Paradigms and Applications*,
Studies in Computational Intelligence 528, DOI: 10.1007/978-3-642-41965-2_9,
© Springer-Verlag Berlin Heidelberg 2014

the cheapest offer for a specific product, booking airline tickets and hotels, attending meetings, participating in decision-making, choosing best educational material, and etc. Among aforementioned services, particularly interesting is the one in which the agents as human representatives participate in group decision-making activities.

In almost all working environments such as business and education, individuals show certain limitations when it comes to efficient problem solving because the problems can occur in various forms, and with greater or less degree of complexity. In order to overcome these limitations, it is common to apply the strategy of joining the group where individuals are able to solve parts of problems that match their competencies and expertise. This is supported by extensive research which has shown that the teams perform better than individuals in a broad range of tasks [1]. A strategy of joining is applied to the agent systems and combining with intelligence, emotions, mood and personality provide basis for simulating processes that occur during group decision-making.

Inside the eLearning environment, agents can be involved in many different types of activities acting as student or professor surrogates. This fact provides a possibility for simulating different group decision-making processes such as evaluation or classification of educational materials. Results of these simulations are very important because they can provide insights necessary for efficient organization or even improvement of the real educational processes. As a representative example, we can use eUniversity system at Faculty of Information Technologies in Mostar which enables students to evaluate and grade every educational material published on the eLearning platform. After that, at the end of the semester, the group of the most successful students can select the best materials on every course. However, instead of students, surrogate agents can be used for simulation of group decision-making aiming to select the best educational materials and thus provide information necessary to determine the quality of each document. Data gathered by these kind of simulations can provide other students with suggestions on the documents that they should study in order to effectively master the course materials.

Notwithstanding the fact that the results of this research were tested on the example of agent group decision-making which aimed to choose the best educational materials, particularly interesting results could be expected in cases where agents represent students on the exam or in cases where agents represent faculty professors in meeting of scientific council. By possessing students' or professors' profiles, it would be easy to simulate group decision-making process.

Analysis of the modern eLearning environments imposes the conclusion that there are a number of reasons why group decision-making can result in failure. Some of these reasons are the disturbed interpersonal relationships between participants, subjectivity, impatience, etc. These factors can be simulated by using a multi-agent system that allows agents with emotions, mood and personality to become human representatives during the meeting and simulate realistic meeting outcome.

Research presented in this chapter analyses group decision-making in eLearning environment during which agents exchange different types of arguments described in [2] trying to convince other participants to adopt their preferences. However, depending on the type of used arguments, the agents experience different kinds of positive and negative emotions that directly affect mood, and therefore the rest of the decision-making process.

The following sections describe the agent model that integrates concepts of patience and experience with personality, emotions and mood in order to constitute a stable and efficient group decision-making process. Introduction of new components is just one more step towards complete simulation of human qualities in order to assign owner's (human) characteristics to an agent which could make them realistic human representatives. The concept of experience is implemented by using Reinforcement learning technique called Q-learning in combination with Self-organizing map while concept of patience is implemented by introducing a Self-regulation coefficient.

The chapter is organized as following. Section 9.2 gives brief overview of the state of the art in applying emotional agents and multi-agent systems in different areas of human live. Section 9.3 describes methods for implementing personality, mood and emotions in the world of agents. Section 9.4 introduces newly proposed components of patience and experience and describes their integration within existing agent model. At the end, Sect. 9.5 describes results of model testing acquired from multi-agent platform. Identified model and multi-agent system shortcomings and suggested directions for future development have been presented in concluding section.

9.2 Literature Review

Development of representative human agents primarily requires an efficient way for mapping human traits into the world of agents. Recent research results have reported successful implementation of methods for mapping some of the most important traits like emotions, mood and personality. The enumerated properties of agents not only have great significance for the individual, but it is particularly interesting their impact on group decision-making. As presented in [3], studies have focused on creating environments for intelligent interaction that can provide support for formal business meetings, tutorials, project meetings, discussion groups and ad-hoc interactions. Context-aware emotion-based agent model presented in [4] ensures that clusters of agents bearing emotion-based features tend to achieve agreements more quickly than those without such features. This model provides a possibility to design intelligent agents with emotional awareness in order to simulate group decision-making processes. Aiming to improve the agent-based negotiation process, authors of [5] have successfully incorporated affective characteristics such as personality, emotion, and mood. Application of the aforementioned characteristics has also been presented in research [6] which introduces a concept of virtual humans with a personality profile and real-time emotions and moods.

The appearance of emotion leads to changes in mood and thus affects agent's state and his future actions. By analyzing the most current research in area of multi-agent systems, we noticed that concepts of patience and experience have been inadequately explored regardless of their great importance for the decision-making processes.

9.3 Agents with Personality, Emotions and Mood

This study is a continuation of the research presented in [4, 5], according to which the structure of agents consists of three layers: knowledge, reasoning and inter-action layer. Figure 9.1 clearly identifies the position of patience and experience, as the newly proposed components, inside the agent structure.

Knowledge layer is the foundation for all other activities of the agent especially for the process of decision-making. Inside this layer, agent stores information about its own condition (preferences, alternatives, goals) and the environment within which it operates.

As with humans, agent reasoning requires the existence of appropriate knowledge that will support this process. Agent reasoning is based on information stored in the knowledge layer that, in addition to reasoning, results with creation of new knowledge. The reasoning layer is realized through several modules: a personality module, a module of emotions, an argumentation module and a decision-making module.

The interaction layer is a kind of interface through which the agent interacts with the environment and other agents who participate in group decision-making process.

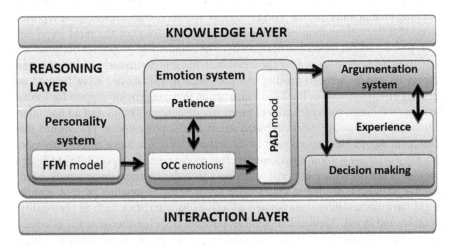

Fig. 9.1 Participant agent architecture

Since the reasoning layer plays the major role in group-decision making process and components of patience and experience have been added to the reasoning layer, the following sections will briefly introduce its modules, while knowledge and interaction layers will not be presented any further.

9.3.1 Agent Personality

Psychology clearly defines interrelationship between personality, emotion and mood. Personality can be regarded as a set of predictable behavior by which people can be recognized and identified [7] or as an individual's pattern of psychological processes, including motives, feelings, thoughts, and other major areas of psychological function [8]. Therefore, personality and individual differences directly affect almost all aspects of human life, starting from the motivation, perception, to cognition.

In the world of agents, personality together with mood and emotional state define the behavior of agents and thus directly affect the process of individual and group decision-making [9].

Personality types are identified by using well-known Five Factor Model (FFM), a model which is accepted as a global framework for the identification of different types of human personalities [10]. FFM model is composed of a set of factors (dimensions) which describe specific personality traits that are rarely changed in the context of time and state. Categorization of individual differences is carried out through the following five dimensions: Openness (curiosity, flexibility, imagination, etc.), Conscientiousness (discipline, organization, reliability, etc.), Extraversion (sociability, penetration, etc.), Amenity (attractiveness, confidence, warmth, etc.) and Neuroticism (anxiety, moodiness, bad temper, etc.). Because of the first letter in the name of each dimension, this model is also called the OCEAN model.

Initially, the identification of five dimensions of personality is performed by using the Big Five Inventory (BFI) model whose results fit with some of the theme based on the FFM model [11]. The theme is a characteristic pattern of personality that occurs as the effect of combining two or more aspects [10].

Based on the FFM model, each personality type defines the following theme categories (styles): leadership style, decision style, learning style, conflict style, sample careers and Holland hexagon. The theme categories which are particularly interesting for the decision-making process are conflict and decision style.

The themes identified in decision style are: Autocrat (N+, O−, A−, C+), Bureaucrat (N−, C+), Diplomat (N−, A, C) and Consensus (N+, E+, A+, C), while conflict style defines following themes: Negotiator (N, E+, A, C−), Aggressor (N+, E+, A−, C+), Submissive (N−, E−, A+, C−) and Avoider (N+, E−, C). Every theme is defined by OCEAN model dimensions where a plus (+) sign indicates a score above 55, a minus (−) sign indicates a score below 45, and a letter without either plus or minus sign indicates a score in the 45–55 range. It is important to emphasize that the theme category also defines a set of arguments that can be used in each of the personality types during argumentation process [12].

9.3.2 Agent Mood

Mood is defined as a feeling or a prolonged emotion that influences the complete state of personality, and thus the perception of the environment [13]. In this context, the mood component is in charge of generating agent's mood based on the intensity of emotions. Compared to other traits, mood can be considered as a trait that is more dynamic than personality, and less dynamic than emotions.

Modeling mood can be achieved by using PAD model which consists of three dimensions: Pleasure (P), Arousal (A) and Dominance (D). These dimensions are fully independent, and can be presented in the PAD 3D space [14, 15]. Thanks to the PAD model, it is possible to map the parameter values of the personality and emotions within the same 3D space. Details of mapping the personality to the PAD mood space are also presented in [16, 17].

When presenting mood, all dimensions are observed through their positive and negative components, where P+ refers to positive pleasure, P− negative pleasure and so on. Positive or negative dimensions of pleasure reflect the emotional state of the agent, excitement is related to physical or mental distress, and dominance reflects a sense of having control. According to model described in [14], we are able to identify the following mood types: exuberant (+P+A+D), bored (+P+A−D), dependent (−P−A+D), disdainful (+P−A+D), relaxed (−P+A−D), anxious (−P−A−D), docile (+P−A−D) and hostile (−P+A+D).

By using five dimensions of the OCEAN model it is possible to present the personality of the agent which can be used for calculation of initial mood. Every personality type has normal emotional state which can be considered as a default emotional state (DES). Mapping the personality to the PAD mood space can be achieved by using Formula 9.1 which is presented in more detail in [14, 16, 17].

$$
\begin{aligned}
&P = O,,,A,N; O,C,E,A,N \in -1,1 \\
&Mood = P,,; P,A,D \in -1,1 \\
&P = 0.59 * A + 0.19 * N + 0.21 * E \\
&A = -0.59 * N + 0.30 * A + 0.15 * O \\
&D = -0.60 * E - 0.32 * A + 0.25 * O
\end{aligned}
\tag{9.1}
$$

Based on emotion intensity, a new point in the PAD space is created that represents the emotional center (EC) which causes the movement of the emotional state (ES). Figure 9.2 shows an example of PAD 3D space with components (marked as points) that are necessary for formation of a new emotional state.

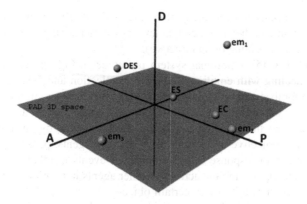

Fig. 9.2 PAD 3D space

The mood ranges from initial points based on personality (DES) to the position where the level of change depends on the type of emotion that caused the change (EC), and slowly returns to its original initial state (ES).

9.3.3 Agent Emotions

Reinhardt et al. [18] argue that agent models inspired by human emotions could provide necessary heuristics in reducing and controlling nondeterminism in decision-making process, flexible cooperation and coordination between agents, efficient human-agent interface that will enhance the interaction between a human and agents.

In psychology, emotions are considered as one of the indispensable components of decision-making process. This consideration is supported by research results which described the influence of emotions on decision-making process, the formation of beliefs and plans, and how people react in certain cases [19–21]. As described in [22], emotions can be presented as episodes of interrelated, synchronized changes in states in response to the evaluation of an external or internal stimulus event as relevant to major concerns of the organism. In the world of agents, emotions are defined as reactions to events, agents or objects, where the reaction is directly dependent on interpretation of certain situations [23].

Agent's emotions are presented by an abbreviated version of the OCC model that is named after its creators Ortony, Clore and Collins. Aforementioned authors presented a total of 22 emotions that can be found in humans [24], after which they presented a shortened version of the original model [23] that contains the following emotions: Joy (because something good happened), Hope (about the possibility of something good happening), Pride (about a self-initiated praiseworthy act), Gratitude (about an other-initiated praiseworthy act), Love (because one finds someone/thing appealing or attractive), Distress (because something bad happened), Fear (about the possibility of something bad happening), Disappointment (because a hoped-for good thing didn't happen), Remorse (about a self-initiated blameworthy act), Anger (about an other-initiated blameworthy act) and Hate (because one finds someone/thing unappealing or unattractive).

As described in [5], emotional system presented in Fig. 9.1 consists of three mechanisms dealing with emotional Appraisal, Selection and Decay.

Emotion appraisal can be viewed as a process in which an emotion arises, and illustrates the way in which a person experiences a particular situation. It certainly does not mean that the situation is just the way perceived by an individual and reflects only his subjective attitude. In the context of agents, phase estimation can generate emotions in response to pleasantness of events in relation to the objectives of the agent, approving the actions of other agents in relation to the standards of behavior, love or hate towards certain objects.

Emotion appraisal well illustrates the differences in emotional reactions to the same event. As an example of situation which should lead to the occurrence of certain emotions, we can use the case in which the agent a_j rejects the request from agent a_i to accept his proposal as the best solution. The intensity of the emotion is calculated using Formula 9.2 which affects process of mapping the emotions to the 3D PAD space. Depending on the mood, certain emotions may not have any impact on decision-making process [6].

Emotion selection provides a mechanism for selecting the most dominant emotions that have arisen as a reaction to some stimulus from the environment (event, action or object). Dominance is determined by value of the difference between intensity and threshold of certain emotions. Considering its direct association with positive and negative reactions, the activation threshold value is obtained as the difference of the two OCEAN model dimensions; Extraversion and Neuroticism [5]. Emotions presented in OCC model are mapped to the PAD space based on the values described in [6]. As an example, emotion of Joy is mapped by using values $P = 0.40$, $A = 0.20$, $D = 0.10$.

As in real life meetings, agents with human traits such as emotion and mood can easily cause group decision-making process fail. One of the reasons is appearance of negative emotions caused by actions of other agents which results in mood change and thus affects the argument selection process. Therefore, patience together with experience can be considered as an adaptation mechanism for group decision-making that will ensure its continuance in critical stages and therefore reduce failures.

9.4 Patience and Experience Within Emotional Agents

The previous part of this work briefly presents personality, mood and emotion, and the basic models which ensure their mapping into the world of agents. The following section describes components of patience and experience together with the results of their application.

9.4.1 Patience Within Emotional Agents

Modern society is increasingly starting to cultivate human qualities such as honesty, sincerity, kindness, etc. which were at risk of decline. However, almost all religions and cultures have always emphasized the importance of a certain trait and this is a trait of patience. The reason why the phrase "Patience is a Virtue" is so common in almost all parts of the world we should probably seek in religious books which say: "Oh you who believe! Persevere in patience and constancy..." (Qur'an, 3:200); "With all humility and gentleness, with patience, bearing with one another in love" (Bible, Ephesians 4:2), "A patient man is better than a warrior, and he who rules his temper, than he who takes a city" (Proverbs 16:32).

In everyday life, patience is considered as the capacity to accept or tolerate trouble or delay without getting angry or upset. However, the astonishing fact is that the significant research in the area of patience has been actualizing only in the last decade. Studies in the psychological literature have typically spoken of patience only as the converse of impatience. Very scarce number of papers deals with the narrow structure and methods of measuring patience. This can be confirmed by the fact that the model for measuring the patience presented in [25] is based on unpublished works that have set the fundamental theoretical structure of patience described in [26, 27]. According to these studies, patience occurs when there is delay of a goal and response depends upon how the person involved evaluates his or her own responsibility in reacting to the delay. In responding to delay, patient people are more serene and compassionate whereas impatient individuals respond with blame and anger.

The aforementioned studies proposed three mechanisms for understanding patient behavior: frustration-aversion, self-regulation and temporal-altruism. Considering patience structure complexity, this research is focused on self-regulation mechanism of patience which is necessary when an individual experiences negative emotions (reaction) and must make appropriate cognitive or behavioral adjustments in order to behave in a patient manner. Self-regulation can be considered as any efforts by the human being to alter any of its own inner states or responses [28]. Although it is reasonable expectation that human beings can control (self-regulate) themselves, the practice has shown that it is not always the case. The management of emotions is perhaps the most challenging, but processes of self-regulation allow for monitoring and evaluation of emotions as well as actions to sustain, enhance, or suppress emotions as desired.

Notion of self-regulation can also be viewed in the context of emotional intelligence which is defined as the ability of an individual to monitor one's own and others' emotions, to discriminate among the positive and negative effects of emotions, and to use emotional information to guide one's thinking and actions [29]. As discussed in [30] self-regulation and emotion regulation are often so intertwined that it is hard to say where one ends and the other begins. When people self-regulate, they are frequently confronted with potentially emotion arousing situations. Learning how self-regulation interfaces with emotion regulation is likely to generate important new insights into both processes. Emotion regulation therefore subsumes the regulation of specific emotions such as anger or fear, along with global mood states, stress, and all kinds of affective responses [30].

In order to integrate a component of patience, we utilized research [31] which distinguishes primary and secondary emotional responses. Primary emotional response is immediate and completely unregulated while secondary emotional response is driven by emotion regulation. The transition between these emotional responds is usually so fast that people hardly notice it and research [32] clearly makes distinction and relation between primary and secondary emotions. According to this research, Joy is considered as primary emotion which can cause secondary emotions of Pride or Relief. However, in this research we only consider emotions defined by the OCC model which are described in previous section.

Patience, as the new component of emotional agents, is incorporated by using two mechanisms. One of them is Self-regulation coefficient (β) which affects the intensity of negative emotions (Em^- is set of negative emotions) as shown in Formula 9.2. The original formula for calculating intensity of emotion, without Self-regulation coefficient, is presented in [5].

$$I_{emotion} = \frac{\sqrt{P^2 + A^2 + D^2}}{\sqrt{3}} * \beta \qquad (9.2)$$

As described in [33], strategies for measuring self-regulation have proliferated and can be divided in three categories: rating-scales, indices derived from behavior and personality inventories. In order to define the value of the Self-regulation coefficient, we used a form of rating-scale category called Self-regulation questionnaire (SRQ) presented in [34]. Based on the SRQ score, described in Formula 9.3, three types of Self-regulation capacities (SRC) have been recognized: High ($\beta = 0.65$), Intermediate ($\beta = 0.75$) and Low ($\beta = 0.85$).

$$\beta = \begin{cases} 1 \text{ if emotion is postive}, em \in Em^+ \\ 0.65 \text{ iff } em \in Em^-; I_{emotion} > activation; SRC = High \\ 0.75 \text{ iff } em \in Em^-; I_{emotion} > activation; SRC = Intermediate \\ 0.85 \text{ iff } em \in Em^-; I_{emotion} > activation; SRC = Low \end{cases} \qquad (9.3)$$

As can be seen from the previous formula, the Self-regulation coefficient is used only in cases of high-intensity negative emotions, and its impact depends on the SRC.

The Self-regulation coefficient is one of two components that are used to implement patience. The other component is based on the research [34] which reports that self-regulation of the secondary emotional response is implemented through monitoring and operating processes. During monitoring process, an agent compares current state with a desired state and after that, an operating process reduces any discrepancies between these two states. In this research we consider emotion of Joy as desired emotional states.

The monitoring process is implemented based on the distance or the correlation between certain emotions presented in [35–37]. Aforementioned researches have investigated the hypothesis that emotions which co-occur frequently within a certain period of time may be relatively accessible to one another; while those that co-occur infrequently may be less accessible, and some of them dealt with the gender differences in emotional correlation. Referring to their earlier research, they constructed a remoteness index of emotional states that yielded shortest paths between different emotions. The remoteness score between two emotions, they argued, provides a quantifiable representation of how much emotion management it takes to move from one emotional state to another. As described in [35], if two emotions have correlation value of 1, then they have remoteness of 0; if they have correlation value of 0 then they have remoteness of 10; and if their correlation value is −1, then they have remoteness of 20. In-between correlations are curvilinear related to remoteness. In other words, a correlation of 0.5 gives remoteness of 3.

When a new emotion arises, a monitoring process is in charge to determine a distance or to find the shortest path between the new and desired emotional state. This is particularly important in cases where an agent has predominantly positive emotions, and suddenly a negative emotion appears.

Output of monitoring process represents input for operating process which, depending on the Self-regulation capacity, uses Algorithm 1.

Algorithm 1. Algorithm for reduction of discrepancies between emotions

```
em_new ← monitor(Em)
if ¬positiveEmotion(em_new) then
    remoteness ← emotionalDistance(em_new, em_desired)
    em_proposed ← operationProcess(em_new, em_desired, remoteness, β)
    Em ← Em ∪ GenerateEmotion(em_proposed)
```

The remoteness index is used inside emotionalDistance for calculating distance between two emotions. Remoteness index describes distances between emotions and Dijkstra algorithm is used to find the shortest path between them. As an example of emotional transition we found that the shortest path from Joy to Anger requires transition over emotion of Pride. By introducing new emotions (for testing purposes) we found that the shortest path from Tranquility to Distress includes transition through emotions of Joy and Fear. After finding the shortest distance between two emotions, operationProcess is in charge of reducing discrepancies between agent's new (em_{new}) and a desired ($em_{desired}$) emotional state. Discrepancies between emotions are realized by using Equation remoteness * SRC which should produce recommended emotional remoteness after suppressing distance with (β) value.

9.4.2 Experience Within Emotional Agents

Life teaches us that experience is a very important factor for success in any field, especially if it comes to negotiating and making decisions. Apart from affecting emotion intensity with Self-regulation coefficient, we believe that the efficiency of group decision-making can be improved by assigning a component of experience to the agents. Since experience requires a lot of learning and practice, we needed to incorporate some mechanism for learning. In this research, an agent learning is provided by using a specific form of Reinforcement Learning called Q-learning [38] in combination with Self-organizing map which enables an agent to decide what action should be taken depending on its current state. Agent's experience in eLearning environment can be used for faster reaching of agreement in group decision-making regarding best educational material.

9.4.2.1 Q-learning

Q-learning can be presented in tabular form, where rows represent the state and the columns represent actions that an agent can perform. Values within the table or Q-values indicate how desirable it is to perform specific actions in relation to the current state. Q-learning enables the analysis of state-action pairs which means that in a particular state, an agent searches for the best action or the action with highest Q-value. Q-learning is realized by using Formula 9.4

$$Q[st, a] \leftarrow Q[st, a] + \alpha(r + \gamma max\, Q\left[st', a'\right] - Q[st, a]) \qquad (9.4)$$

where Q[st, a] represents the Q-value for the current state-action pair, Q[st', a'] represents the Q-value for the next state-action pair, r is the reward, α is the learning rate, γ is the discount factor and therefore r + γ max/Q[st/, a/] represents the actual current reward plus the discounted estimated future value for being at the state st and doing action a.

9.4.2.2 Self-Organizing Map

Placing Q-values in tabular form shows very good results in cases where the number of state-action pairs is not very large. However, if the agent is stationed in a dynamic and nondeterministic environment, then there can be certain performance and implementation issues. As a possible solution to the aforementioned problems, research results [39–41] have imposed the usage of special kind of neural networks called Self-organizing map (SOM) or Kohonen map.

SOM [42] is a type of neural network that learns to classify input vectors according to how they are grouped in the input space and therefore is able to reduce a large input space to a fixed-size, usually two-dimensional output space. These excellent SOM properties are very suitable for data classification and are based on the learning rule by which vectors that are similar to each other in multidimensional space should be similar in two-dimensional space.

SOM neural network usually consists of input layer used only for accepting input data and output layer which contains network of neurons. Input neurons are connected to all output neurons. After the vector is presented to the input layer, self-organizing network performs entire set of phases in order to produce desired formation (topological structure) and thus ensure learning [43–45]. The first step of the process of network formation is a competition in which the neurons in the output layer compete for each sample presented to the network. To represent the process of cooperation, we observed n-dimensional vector of input data $\vec{x} = [x_1, x_2, \ldots x_n]^T$ and weights vector $\vec{w_j} = [w_{j1}, w_{j2}, \ldots w_{jn}]^T$. The value of j in the weight vector ranges from 0 to m where m is the number of output neurons, and therefore SOM will have n by m number of network connection between input and output layer. The goal is to find the cooperative members (neuron) that best fit the relation between vectors $\vec{x}$ and $\vec{w_j}$. Minimal Euclidean distance $\left\| \vec{x} - \vec{w_j} \right\|$ will determine the winning neuron which is obliged to make adjustment in order to narrow distance between input vector and its weights. Since all neurons have a set of neighbors, winning neurons will cause neighbor weights to adjust. The further the neighbor is from the winner, the smaller its weight change. Furthermore, as training goes on, the neighborhood gradually shrinks.

As already pointed out, instead of Q-table, a SOM neural network is used to store the Q-values. For the purposes of training SOM, we have isolated the required values from 50 simulations in which four agents with different characteristics have participated. During training phase SOM associates each neuron of the map to an emotional state of the agent, personality type of the agent with which it is currently negotiating, type of argument used for negotiation, reward (used to update Q value) and the Q-value. In other words, the input vector can be formalized as follows $\vec{x} = \left[a_{i_{emotionalState}}, a_{j_{personality}}, a_{i_{argument}}, Q \right]^T$.

Values used to represent emotional state, personality and argument are in the range 0–1, for example, the most positive emotion has value 0.1 while the most negative emotion has value 0.9. Reward can have one of three values: +1 if agent a_j accepted request, 0 if agent a_j is analyzing request, and −1 if agent a_j rejects the request.

After training, the SOM neural network could be used for group decision-making. An agent consults SOM network during every stage of negotiations or before sending any request to other agents aiming to select the best argument. The process of consulting SOM network requires an agent to send a vector with identical dimensionality as it was during the training phase. Instead of actual Q-values, an agent sends 1 in order to discover neurons that best suit current emotional state and personality of the agent which is currently negotiating with. After finding neurons that best suit the presented vector, the agent uses a type of argument which will be used for sending the request. The response to the sent request allows modification of the Q-value which ensures continuation of learning process.

9.5 Model Evaluation

At the beginning of this research, we had the assumption that the components of patience and experience will contribute to the faster reaching of an agreement and more stable group decision-making. In order to test this assumption with emotional agents, we used multi-agent system developed in GECAD,[1] where we made a milder adjustment in order to implement newly proposed components of patience and experience. In the simulation environment, agents represented surrogates of the students during the meeting where they needed to choose the best course material. Since the eUniversity system at Faculty of Information Technologies allows students to grade every course material published on eLearning platform, the easiest way to solve the aforementioned dilemma would be to select a material with the highest overall score assigned by all students enrolled in course. However, in the simulation environment selection process was left to the group of the most successful students where they had to consider different characteristics of the material such as: Authors competence, Technical accuracy of content, Usefulness, Illustrations and Overall rating. Average grades assigned by students enrolled in course represent the basis for decision process, and are shown in Table 9.1.

In addition, each of the agents (surrogate—representative of the most successful student) has different preferences when it comes to selecting the best materials and those preferences are shown in Table 9.2. Preference values are in range 0–1.

The method of decision-making used is known as Multi-Criteria Decision Making (MCDM). Based on the established criteria, this decision making process allows to make a selection of alternatives that best match the goals and desires of an individual [46].

In order to fully evaluate the application of patience and experience in group decision-making, we performed 10 simulations (SM) for different agents' profiles presented in Table 9.3. During the simulation, we observed the occurrence and intensity of negative emotions as well as their impact on the number of exchanged arguments in the context of reaching an agreement. The previous research [5] has reported that the combination of avoider and submissive agents results in a reduced number of exchanged arguments while the number of exchanged arguments tends to increase with the combination of negotiator and aggressor agents. Therefore, as shown in Table 9.3, the majority of the agents in the simulation had personality of negotiator and aggressor with different values of Self-regulation capacity.

During the analysis, each of the negative emotions has been assigned a certain weight in the range of 0.5–0.9, and we summarized their values in each of the simulations. As presented in Fig. 9.3, during 10 simulations, agents bearing patience and experience features tend to achieve agreements more quickly with less negative emotions than those without such features.

---

[1] GECAD (Knowledge Engineering and Decision Support Research Center http://www.gecad.isep.ipp.pt/), Porto.

Table 9.1 Average material grades assigned by students

Criteria	Course material 1	Course material 2	Course material 3	Course material 4
Authors competence	55	73	60	82
Technical accuracy of content	80	65	70	58
Usefulness	41	82	50	69
Illustrations	70	66	51	70
Overall rating	65	78	82	59

Table 9.2 Agents' preferences

Criteria	Authors competence	Technical accuracy of content	Usefulness	Illustrations	Overall rating
Agent 1	0.46	0.01	0.35	0.39	0.06
Agent 2	0.40	0.53	0.13	0.05	0.43
Agent 3	0.15	0.44	0.01	0.10	0.25
Agent 4	0.43	0.40	0.39	0.27	0.51

Table 9.3 Agent profiles

Simulation	Features	Agent 1	Agent 2	Agent 3	Agent 4
SM1	Personality	Negotiator	Negotiator	Negotiator	Aggressor
	SRC	High	Intermediate	Low	High
SM2	Personality	Negotiator	Negotiator	Negotiator	Aggressor
	SRC	High	High	Low	Intermediate
SM3	Personality	Negotiator	Negotiator	Negotiator	Aggressor
	SRC	Low	Low	Intermediate	Low
SM4	Personality	Negotiator	Negotiator	Negotiator	Submissive
	SRC	Intermediate	Intermediate	Low	Intermediate
SM5	Personality	Negotiator	Negotiator	Negotiator	Avoider
	SRC	Intermediate	Intermediate	Low	Low
SM6	Personality	Aggressor	Aggressor	Aggressor	Negotiator
	SRC	High	Intermediate	Low	High
SM7	Personality	Aggressor	Aggressor	Aggressor	Negotiator
	SRC	High	High	Low	Intermediate
SM8	Personality	Aggressor	Aggressor	Aggressor	Negotiator
	SRC	Low	Low	Intermediate	Low
SM9	Personality	Aggressor	Aggressor	Aggressor	Submissive
	SRC	Intermediate	Intermediate	Low	Low
SM10	Personality	Aggressor	Aggressor	Aggressor	Avoider
	SRC	Intermediate	Intermediate	Low	Low

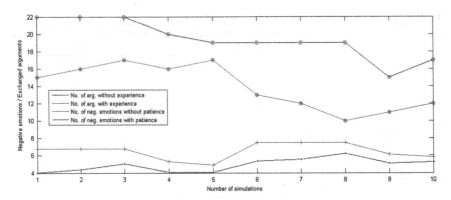

Fig. 9.3 Effect of applying patience and experience on group decision-making with emotional agents

Compared to [47], results of the analysis suggest a few very interesting conclusions. First of all, component of experience and patience have the greatest effect in case where the majority of agents in the group had negotiating personality in combination with an agent who is described as the aggressor. The previous conclusion is made based on the level of negative emotions and the lowest number of exchanged arguments during group decision-making. In this case, all the participants had a high or medium level value of self-regulation coefficient.

The proposed components achieve the smallest influence in group decision-making in case where the majority of agents have a negotiating personality together with an agent declared as the avoider. In this case most of the agents had intermediate or low level value of self-regulation coefficient. The reason for the smallest influence could be found in the fact that avoider personality itself does not support any form of confrontation or aggression, which certainly contributed to reducing tensions in group decision-making. It is also interesting to emphasize that the components of patience and experience have the biggest influence concerning the number of exchanged arguments in case when the majority of aggressor and negotiator agents with low self-regulation coefficient value is found in the group.

Therefore, research results support the fact that the components of patience and experience ensure positive effects and can be considered as desirable and implementable components in agent structure, especially concerning agents which are involved in group decision-making processes.

9.6 Conclusion and Future Work

The most significant challenges in the implementation of the newly proposed components are related to the integration of Reinforcement learning and Self-organizing neural networks in the context of emotional agents. The greatest amount of time was spent on the identification and extraction of attribute values that ensure adequate data classification and agent learning. This is also directly related to one of the greatest shortcomings of the multi-agent system used for model testing, and that is the number and type of arguments that agents can use during group decision-making. A small number of available arguments prevent long negotiation processes to occur, which reduces the amount of high-quality data that could be used for feeding Self-organizing network and ensure more realistic growth of experience. That is why we believe that the implementation of a different argumentation model could significantly improve multi-agent system used for model testing and provide more efficient insights in group decision-making process.

In addition, we should also mention that a significant part of the research carried out in determining the value of Self-regulation coefficient. In fact, we sought to use one coefficient in a two-stage process: lowering the intensity of negative emotions and reducing negative emotional transition. The proposed values are acceptable in the context of research beginnings, but certainly there is room for refinement.

Another major deficiency of the current model is the existence of two questionnaires that need to be fulfilled in order to assign personality and determine the level of self-regulation for each agent. If we take into account that the decision-making process involved four agents, this means that each group decision-making session requires filling eight questionnaires. During the model testing, aforementioned values were predefined and automatically set, but this certainly represents a significant deficiency that future versions of the tested multi-agent system should overcome.

Regardless of the aforementioned multi-agent systems shortcomings and suggested directions for future development, research results clearly confirmed that the introduction of the Self-regulation coefficient and experience directly affects the reduction of negative emotions and number of exchanged arguments. As a result, newly proposed components also affected the speed of reaching an agreement within the group which represents a reason enough for their improvement and further research. Also, we believe that significant research results could be made by analyzing the relationship between self-regulation and the Neuroticism in order to affect emotional transitions.

Acknowledgments Special thanks go to Carlos Ramos (Vice president of the Polytechnic Institute of Porto) and Goreti Marreiros (Knowledge Engineering and Decision Support Research Center-GECAD) on their generous assistance and provided opportunity to collaborate with them.

References

1. Beyerlein, M.M., Johnson, D.A., Beyerlein, S.T.: Advances in Interdisciplinary Studies of Work Teams. JAI, Greenwich (1997)
2. Sarit, K., Katia, S., Amir, E.: Reaching agreements through argumentation: A logical model and implementation. Artif. Intell. **104**(1–2), 1–69 (1998)

3. Tate, A., Chen-Burger, Y.H., Dalton, J., Potter, S., Richardson, D., Stader, J., Wickler, G., Bankier, I., Walton, C., Williams, P.: I-room: A virtual space for intelligent interaction. IEEE Intell. Syst. **25**(4), 62–71 (2010)
4. Marreiros, G., Santos, R., Ramos, C., Neves, J.: Context-aware emotion-based model for group decision making. IEEE Intell. Syst. **25**(2), 31–39 (2010)
5. Santos, R., Marreiros, G., Ramos, C., Neves, J., Bulas-Cruz, J.: Personality, emotion, and mood in agent-based group decision making. IEEE Intell. Syst. **26**(6), 58–66 (2011)
6. Patrick, G.: ALMA: A layered model of affect. In: 4th International Joint Conference on Autonomous Agents and Multiagent Systems; AAMAS '05, pp. 29–36 (2005). ISBN: 1-59593-093-0
7. Richard, M.R.: Theories of Personality. Wadsworth Publishing Company, Belmont (2003). ISBN 978-0534619886
8. Mayer, J.D.: A Classification of DSM-IV-TR Mental Disorders According to their Relation to the Personality System. Comprehensive handbook of personality and psychopathology (CHOPP). vol. 1, Wiley, Hoboken (2006). ISBN: 0471739138
9. Arjan, E., Sumedha, K., Nadia, M-T.: A Model for Personality and Emotion Simulation. Lecture Notes in Computer Science, vol. 2773, pp 453–461 (2003). ISBN: 978-3-540-40803-1
10. Howard, P.J., Howard, J.M.: The big five quickstart: An introduction to the five-factor model of personality for human resource professionals. Education Resource Information Center (1995)
11. Oliver, P.J., Sanjay, S.: The big five trait taxonomy: History, measurement, and theoretical perspectives. In: John, O.P., Robins, R.W., Pervin, L.A. (eds.) Handbook of Personality: Theory and Research, pp. 102–138. The Guilford press, New York (1999)
12. Ricardo, S., Goreti, M., Carlos, R., José, N., José, B.-C.: Using personality types to support argumentation. In: 6th International Conference on Argumentation in Multi-Agent Systems, ArgMAS'09, pp. 292–304. Springer, Berlin (2010). ISBN:3-642-12804-1
13. Anita, V.C.: Psychology of Moods. Nova Publishers, Hauppauge (2005). ISBN 978-1594543098
14. Mehrabian, A.: Analysis of the big-five personality factors in terms of the PAD temperament model. Aust. J. Psychol. **48**, 86–92 (1996)
15. Frijda, N.H.: The Laws of Emotion. Lawrence Erlbaum Associates, (2006). ISBN: 978-0805825978
16. Diana, A., Javier, V., Francisco, J.P.: Generation and visualization of emotional states in virtual characters. Comput. Anim. Virtual Worlds. **19**(3–4), 259–270 (2008)
17. Van der Lei, K.: A review of the current state of emotion modeling for virtual agents. In: 14th Twente Student Conference on IT, vol. 14. University of Twente, Enschede (2011)
18. Reinhardt, D., Levi, P., Meyer, J.-J.C.: Emotions as heuristics in multi-agent systems. In: Proceedings of the 1st Workshop on Emotion and Computing—Current Research and Future Impact (2006)
19. Grossberg, S., Gutowski, W.: Neural dynamics of decision making under risk. Psychol. Rev. **94**, 300 (1987)
20. Damásio, A.R.: Descartes' Error: Emotion, Reason and the Human Brain. G.P. Putnam's Sons, New York City (1994). ISBN 978-0-399-13894-2
21. De Sousa, R.: Why think? Evolution and the Rational Mind. Oxford University Press, Oxford (2002). ISBN 978-0195189858
22. Klaus, R.S.: What are emotions? And how can they be measured? Social Sci. Inf. **44**(4), 695–729 (2005)
23. Andrew, O.: On making believable emotional agents believable. In: Trappl, R., Petta, P., Payr, S. (eds.) Emotions in Humans and Artefacts, pp. 189–212. MIT Press, Cambridge (2003)
24. Andrew, O., Gerald, L.C., Allan, C.: The Cognitive Structure of Emotions. Cambridge University Press, Cambridge (1990). ISBN 978-0521386647

25. Dudley, K.C.: Empirical development of a scale of patience. Dissertation submitted to the College of Human Resources and Education, West Virginia University (2003)
26. Blount S., Janicik, G.A.: Comparing Social Accounts of Patience and Impatience. Unpublished manuscript, University of Chicago (1999)
27. Blount S., Janicik, G.A.: What makes us Patient? The Role of Emotion in Socio-Temporal Evaluation. Unpublished manuscript, University of Chicago (2000)
28. Baumeister, R.F., Vohs, K.D.: Handbook of Self-Regulation: Research, Theory, and Applications. Guilford Press, New York (2004). ISBN 978-1572309913
29. Salovey, P., Mayer, J.: Emotional Intelligence. Imagination, Cognition and Personality, vol. 9, pp. 185–211. Baywood Publishing, New York (1990)
30. Sander, L.K., Lotte, F.D., Gal, S.: The self-regulation of emotion. In: Boekaerts, M., Pintrich, P.R., Zeidner, M. (eds.) Handbook of Self-Regulation. The Guilford Press, New York (2010). ISBN 978-1-60623-948-3
31. Lazarus, R.S.: Progress on a cognitive-motivational-relational theory of emotion. Am. Psychol. **46**(8), 819–834 (1991)
32. Parrott, W.: Emotions in Social Psychology. Psychology Press, London (2001). ISBN 9780863776823
33. Rick, H.H., Erin, K.B.: Measurement and modeling of self-regulation: Is standardization a reasonable goal? National Research Council Workshop on Advancing Social Science Theory (2010)
34. Charles, S.C., Michael, F.S.: On the Self-Regulation of Behavior. Cambridge University Press, Cambridge (2001). ISBN 978-0521000994
35. Kathryn, L.J., David, R.H.: Sociological realms of emotional experience. Am. J. Sociol. **109**(5), 1109–1136 (2004)
36. Kathryn, L.: Emotional segues and the management of emotion by women and men. Soc. Forces **87**(2), 911–936 (2008)
37. Morgan, R.I., David, R.H.: Structure of emotions. Social Psychol. **51**(1), 19–31 (1988)
38. Andrew, G.B., Richard, S.S.: Reinforcement learning: An introduction. In: Sutton, J., Barto, A.G. (eds.) A Bradford Book. MIT Press, Cambridge (1998). ISBN 9780262193986
39. Smith, A.J.: Applications of the self-organizing map to reinforcement learning. J. Neural Netw.: New Dev. Self-Organ. Maps **15**(8-9), 1107–1124 (2002)
40. Touzet, C.: Neural reinforcement learning for behavior synthesis. Robot. Auton. Syst. **22**(3–4), 251–281 (1997)
41. Touzet, C.: Modeling and simulation of elementary robot behaviors using associative memories. Int. J. Adv. Robot. Syst. (2006). ISBN: 1729-8806
42. Kohonen, T.: Self-Organizing Maps, 3rd edn. Springer, Berlin (2001). ISBN 978-3540679219
43. Ben, K., Patrick, S.: An Introduction to Neural Networks. Lecture Notes (1996)
44. Tom, M.: Machine Learning. McGraw-Hill Science/Engineering/Math, NY (1997). ISBN 978-0070428072
45. Taylor, J.G.: Mathematical Approaches to Neural Networks. Elsevier, Amsterdam (1993). ISBN 9780080887395
46. Evangelos, T.: Multi-Criteria Decision Making Methods: A Comparative Study. Applied optimization, vol. 44. Springer, Berlin (2000). ISBN: 978-0792366072
47. Music, D.: Patience in group decision-making with emotional agents. In: Trends in Practical Applications of Agents and Multiagent Systems, PAAMS 2013, vol. 221, Springer, Berlin (2013). ISBN: 978-3-319-00562-1

About the Editors

Mirjana Ivanović is holding a position of full-time professor since 2002 in the Faculty of Sciences, University of Novi Sad, Serbia. She is a head of Chair of Computer Science and a member of University Council for informatics. She authored/co-authored 13 textbooks and more than 235 research papers on multi-agent systems, e-learning and web-based learning, software engineering education, intelligent techniques (CBR, data, and web mining), most of which are published in international journals and international conferences. She is/was a member of Program Committees of more than 120 international Conferences and is the Editor-in-Chief of *Computer Science and Information Systems* journal.

Lakhmi C. Jain is with the Faculty of Education, Science, Technology, and Mathematics at the University of Canberra, Australia and University of South Australia, Australia. He is a Fellow of the Institution of Engineers Australia.

Dr. Jain founded the KES International for providing a professional community the opportunities for publications, knowledge exchange, cooperation, and teaming. Involving around 5,000 researchers drawn from universities and companies worldwide, KES facilitates international cooperation and generate synergy in teaching and research. KES regularly provides networking opportunities for professional community through one of the largest conferences of its kind in the area of KES. www.kesinternational.org.

His interests focus on the artificial intelligence paradigms and their applications in complex systems, security, e-education, e-healthcare, unmanned air vehicles, and intelligent agents.

M. Ivanović and L. C. Jain (eds.), *E-Learning Paradigms and Applications*,
Studies in Computational Intelligence 528, DOI: 10.1007/978-3-642-41965-2,
© Springer-Verlag Berlin Heidelberg 2014

Author Index

M. Ivanović and L. C. Jain (eds.), *E-Learning Paradigms and Applications*,
Studies in Computational Intelligence 528, DOI: 10.1007/978-3-642-41965-2,
© Springer-Verlag Berlin Heidelberg 2014

Printed in the United States
By Bookmasters